Bicycling Science

Bicycling Science Ergonomics
and
Mechanics

Frank Rowland Whitt
and
David Gordon Wilson

The MIT Press Cambridge, Massachusetts, and London, England

This book was set in IBM Composer Univers,
and printed and bound by The Colonial Press, Inc.
in the United States of America

Second printing, first MIT Press paperback edition, 1976

Library of Congress Cataloging in Publication Data

Whitt, Frank Rowland.
    Bicycling science: ergonomics and mechanics.

    Includes bibliographical references.
    1. Bicycles and tricycles—Dynamics. 2. Man-machine
systems. I. Wilson, David Gordon, 1928-joint author.
II. Title.
TL410.W48 629.22'72'015313 74-5165
ISBN 0-262-23068-2

This book is dedicated to all those men and women whose efforts have produced the bicycle of today—the simplest, quietest, most efficient and least lethal of modern vehicles. In particular, we wish to pay our respects to the memory of Dr. Paul Dudley White, world-reknowned heart specialist, who did more than any other person to reawaken appreciation for the bicycle as a contributor to health.

It is terrifying to think how much research is needed to determine the truth of even the most unimportant fact.

—Stendhal

Contents

Foreword

This book is intended to be of interest to all mechanically inquisitive bicyclists, as well as to teachers of elementary mechanics or physiology, and to engineers and others working on approaches to lessen our dependence on high-energy-consumption transportation.

It should also show other bicycle users how much scientific work has been put into the exploration of just a few of the less obvious aspects of the use of a bicycle.

The intense interest in bicycles during the Victorian "boom" period of the 1890s gave rise to much detailed literature on the mechanical side of the machine itself. A classic book written by A. Sharp in 1896 is a very good example of the best of such technical writing. R. P. Scott and C. Bourlet wrote other good books of this period. These books appear to have been the last of their kind. It seems that after 1896 competent authors turned their attentions to the "horseless carriage." Only a few appear to have continued to write on bicycling developments, and their material was then published only infrequently in those periodicals on bicycling which survived into the twentieth century.

Technical advances have been made since 1896, but widespread specialization has occurred. It is now necessary to search scientific and engineering journals, seemingly but distantly connected with cycling matters, to find technical information of the type written about by Sharp in his comprehensive book.

The history of modern road vehicles shows that their evolution has been subscribed to by many types of inventors, manufacturers and businessmen whose opinions on the best methods of approach to production could be at considerable variance. It is not altogether surprising that some products of this combination appeared to have been conceived with but little attention having been paid to well-established scientific principles. Some of these particular products, among which were bicycles, were therefore doomed to failure.

One of the aims of this book is to provide the type of information which could enable some potential future failures to be avoided. There are some basic principles associated with bicycle motion which, unlike the detailed design of the machine, do not change with time or fashion and are the same for a rider of a veteran "manumotor" or of a modern bicycle. The power producible by a human and the laws governing the forces with which wind and road conditions oppose machine motion are unalterable by time or man. A knowledge of these basic facts can assist, at least, in making a sound prediction of the limits to any improved performance which a change in bicycle design could be expected to give a rider.

This is a book describing the measurements and experiments which have been made in connection with the above basic principles, and some of the designs which have resulted from their application. The basic principles treated here are concerned with dynamics rather than statics. We start with energy requirements for transportation, and continue with the study of the power producible by humans in various ways. Then we review the natural forces opposing motion and the applied braking forces; cooling effects on the rider and steering and stability. Some unusual applications of "people-power" to transportation, and a look at new developments, complete this book.

The basic text was prepared by Whitt partly as a compilation of articles written over the years as a contributor to *Cycle Touring* (Cyclists Touring Club), *Bicycling!,* and other bicycling magazines. Also as a result of many years of experience in scientific research work recorded in many publications. The text was edited by Wilson who added the results of research and design studies carried out under his supervision at the Massachusetts Institute of Technology, and some details of the results of an international design competition which he organized on man-powered land transport in 1969-1970.

Bicycling is experiencing a wave of renewed

popularity of a magnitude that has amazed even the enthusiasts. Much of the new growth will be on stony ground, and will wither. But with the simultaneous imposition of ever-more rigorous controls on pollution in our cities and the growing shortages in the supply of energy for motorized transportation, it seems certain that there will be an increasing incentive to find ways to allow people to move themselves about through at least short distances without the aid of 3,000 pounds [1360 kg] of automotive machinery. The bicycle presents itself as an even more efficient user of transportation energy than the dolphin, and its use—or the use of something like it—is bound to increase. We hope that this book will enable bicyclists, old hands and newcomers, to understand their pursuit better, and engineers and inventors to change the future more wisely.

Note on units

This book was originally written using British (Imperial) engineering units—feet, pounds, degrees Fahrenheit, and so forth. We believe that most of our readers still feel more comfortable with these units. However, we have given equivalent S. I. units in square brackets in the text and in most of the graphs. To duplicate the tables would, we felt, have been unwarranted, and they remain in British units.

Note on "he"

Bicycling had a very significant role in the beginnings of women's emancipation in the 1880s onwards, and it has continued to be a popular women's sport and recreation, as well as a means of transportation. We have frequently referred to bicyclists as males here only because of the awkwardness of using such devices as "his or hers" continually.

Acknowledgments

Many individuals and organizations have helped to make this book possible.

Those who have given permission to reproduce copyrighted material are acknowledged on the particular illustrations and are remembered here with appreciation.

William A. Bush, Evelyn Beaumont, Vaughan Thomas, Derek Roberts, and David E. Twitchett and other members of the Southern Veteran Cycle Club, England, and the Camden Historical Society have loaned items of equipment for testing or pieces of historical literature for study and copying.

H. John Way, editor of *Cycle Touring* (the magazine of the Cyclists' Touring Club), has allowed the use of a considerable number of articles contributed to that magazine over the years by the senior author.

Frederick Arthur DeLong, technical editor of *Bicycling!*, has exchanged information with both authors for many years and acts as an efficient and cordial clearing house of technical information on bicycling in the United States.

Anna Piccolo, secretary at Massachusetts Institute of Technology, has cheerfully and competently typed the manuscript and tables and organized references and the index, largely at times when she would otherwise have been able to relax from a too-heavy workload.

To all of these generous people the authors express their very sincere gratitude and appreciation.

Part 1

Human power

1 Power needed for land locomotion

All things upon the surface of the earth, to varying degrees, move. Climatic changes, due to the movement of the earth itself, cause water to evaporate and fall as rain, which in turn, with the help of gravity, wears away land masses. Winds blow and rivers flow, moving large quantities of matter. In order to survive, living species like animals and man had to develop controllable movement, independent of gravitational and fluid forces which are the usual basis for movement of inanimate objects. The animal world developed "lever systems," which pushed against the ground in various ways from crawling, as do snakes, through bounding, like rabbits, to walking, as practiced by man, which in some ways is like the rolling of a spoked but rimless wheel. With the adoption of the wheel, yet another lever mechanism for movement, came the chance, now fully exploited, of using a separate inanimate source of power other than that of the muscles of the moving creature. Steam, internal-combustion-engine, and electric vehicles rapidly appeared when lightweight engines of adequate power had been produced.

The bicycle is only one of the many man-developed lever systems for land transport, but it is the sole remaining type that has a limited propulsive power. All other wheeled vehicles have, in general, been fitted with driving units of progressively increased power. In ancient times teams of horses or cattle succeeded single draught animals. The urge for more power and speed seems ever present in the activities of man with, as yet, no sign of it being satisfied.

Power needed by animals or wheels to cause movement

The relative power needed to move a vehicle or animal against ground resistance by various means is shown in Figure 1.1. At speeds of a few miles per hour these sliding, crawling, leaping, or rolling mo-

tions absorb almost all the power exerted by the
subject, so that wind resistance can be neglected
for purposes of approximate comparison. At higher
speeds, the resistance to motion due to air friction assumes a dominant role and obscures the
more fundamental difference between wheel motion and other systems of movement based on
leverage.

Lever systems are intrinsically efficient, and
Figure 1.1 (which includes data from Bekker[1])
shows that Nature, in developing walking for
man's progression, has given him a system more
economical in energy use than that employed by
many other animals. Nature has also arranged for
her lever systems to be adjusted automatically
according to the resistance encountered. The
stride of the walker changes, for instance, according to the gradient. In this respect, the rider of a
bicycle is at a disadvantage, because bicycle gearing that automatically adjusts to give an optimum
pedaling rate is not yet available. Such a device
would have advantages when a high power output
over varying conditions is wanted. Modern multi-
geared bicycles can approximate, if skillfully used,
an automatic infinitely variable gear. For low
power output, such as is needed for low speeds,
the combination of foot pressure and crank revolution rate is not critical.

**Power required by
bicycles compared
with that for other
vehicles**

The bicycle and rider, in common with most other
wheeled vehicles, can move over hard smooth
surfaces at speeds at which air resistance is significant, that is, at speeds greater than the 5 mile/h
[2.2 m/sec] upper limit of Figure 1.1. The sum
total of wind resistance, ground-movement resistance, and machinery friction decides the rate of
progress for a given power input to a vehicle. These
resistances have been studied carefully over a long
period for the commonly used machines, such as
those using pneumatic tires on pavement and steel
wheels on steel rails.

Graphs showing how the individual resistances
contribute to the total for bicycles, railway trains,

Figure 1.1
Propulsion-power require-
ments for animals and
vehicles. Some of the data
are from Bekker, reference 1.

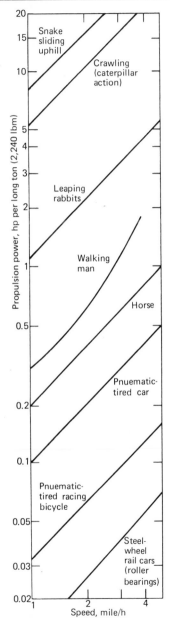

and automobiles are given in Figures 1.2, 1.3, and
1.4. In each case typical examples of vehicles without special streamlining treatment have been chosen
in order to bring out reasonable comparisons. The
tricycle has been included because it shows the
incremental effort needed for propulsion—up to
10 percent above that for the bicycle—as can be
deduced from the times achieved in races by
riders of the two types. Published information
concerning the powers of "mopeds" (low-power
motorcycles) and their performances is included
in Table 1.1.

The basis for these data will be explained in
more detail later. Our present purpose in compar-
ing these various means of locomotion is to place
the bicycle in a relationship to other common
road vehicles. Some relative power requirements
are shown in Figure 1.5. Table 1.2 shows that of
all the vehicles, bicycles are impeded the most
by winds. Figure 1.3 shows that a feature of
modern automobiles is the relatively high power
absorbed by tires. In contrast, railway trains are
hardly affected by wind resistance below 40 mile/h
[17.9 m/sec]. With regard to the propulsion power
required per unit weight, the bicyclist can be seen
to need far less than the walker at low speeds.
This advantage will be examined more closely
later.

Animal power

The power available for propelling a bicycle is
limited to that of the rider. Let us study how
human muscle power compares with that of other
living things with similar muscle equipment.
For thousands of years, and even today in the
less-developed parts of the world, horses, cattle,
dogs and humans have been harnessed to machines
to provide power to turn mills, lift water buckets,
and do other domestic tasks needing power. When
the steam engine was invented, it was necessary
to have handy a comparison between its power
and that of a familiar source. Experiment showed
that a big horse could maintain for long periods
a rate-of-lifting power equal to that of raising

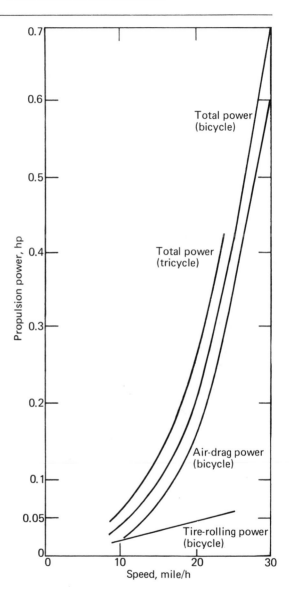

Figure 1.2
Propulsion-power require-
ments for bicycles and
tricycles.

Figure 1.3
Propulsion-power require-
ments for automobiles.
Automobile mass, 2,240
lbm; frontal area, 20 sq ft.

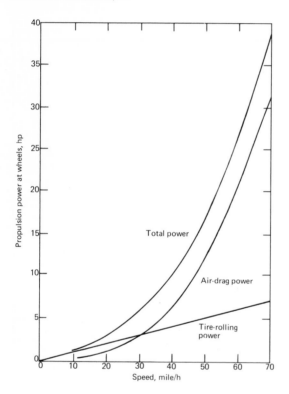

Figure 1.4
Freight-train power requirements. Data from Trautwine,
reference 2, p. 1058.

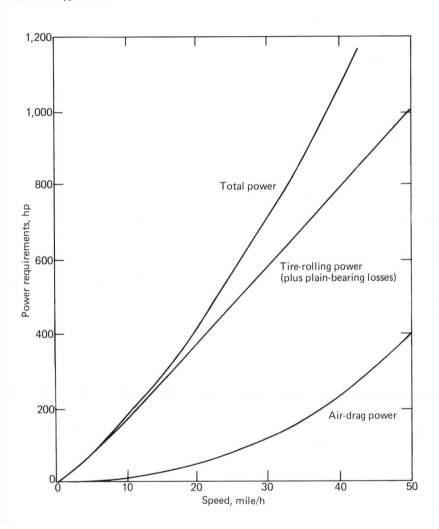

Figure 1.5
Propulsion-power require-
ments over a range of
speeds. Data from
Table 1.3.

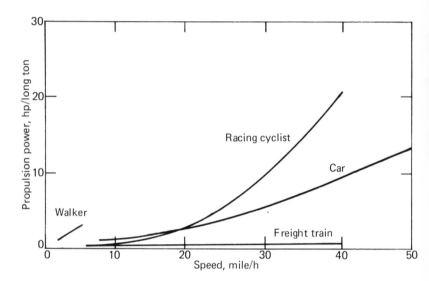

Table 1.1. Power to propel mopeds.

Make	Engine power, hp	Engine speed, rpm	Moped weight, lbm	Rider, lbm	Level-road max. speed (mile/h)	Wheel diam., in.
Powell[a]	1.05	3,500			26	
Mobylette	1.35	3,400	75	200	30	26 (approx)
Magneet[b]	1.6	4,700	115	200	33	
Raleigh	1.4	4,500	77	182	26	16

Sources:
[a]*Cycling,* 9 July 1958, p. 24.
[b]*Cycling,* 27 June 1957, p. 537.

Table 1.2. Estimated forces opposing the motion of various vehicles on smooth surfaces in still air (typical cases).

Vehicle and weight	Origin of force	Resisting force at various speeds, lbf			
		5 mile/h	10 mile/h	20 mile/h	40 mile/h
Man walking 150 lbm	Wind	0.2			
	Rolling	13.0			
	Total	13.2			
Cyclist 170 lbm (racing type)	Wind	0.2	0.8	3.2	12.8
	Rolling	0.9	0.9	0.9	0.9
	Total	1.1	1.7	4.1	13.7
Auto 2240 lbm	Wind	0.9	3.5	14.0	56.0
	Rolling	37.0	37.0	37.0	37.0
	Total	37.9	40.5	51.0	93.0
Freight train 1500 tons	Wind	35	140	560	2,250
	Rolling	7,500	7,500	7,500	7,500
	Total	7,535	7,640	8,060	9,750

33,000 pounds [14,968.5 kg] one foot [0.3048 m] in one minute. This figure (and its equivalents of reduced weight with increased height or decreased time combinations) came to be universally accepted as the "horsepower." Average horses, could, in fact, work at a greater rate but only for briefer periods which were not useful. Trautwine expands upon the relationships between total output per day and rate of output.[2]

Other information relating peak power output to the duration of effort is given in Table 1.3 and Figure 1.6. It seems that a man tends to adjust his power output to rather less than one tenth of a horsepower [74.6 watts] if he intends to work for other than very short periods and is not engaged in competition. Engineering textbooks giving calculations concerning the motion of bicyclists have for over 70 years supposed that 0.1 hp [74.6 watts] was a reasonable figure to use for a man cycling under continued and level conditions. Detailed analysis of this phenomenon is given in succeeding chapters showing that such a

Table 1.3. Power outputs of horse and man.

Subject and conditions	Period	Power, hp
Horse galloping (27 mile/h)[a]	2 min	4 - 5 (per ton)
Towing barge (2½ mile/h)[b]	10 h	0.67
Man towing barge (1½ to 3 mile/h)[b]	10 h	0.11
Turning winch[b]	10 h	0.058
Treadmill[b]	10 h	0.081
Climbing staircase[c]	8 h	0.12
Turning winch[c]	2 min	0.51

Sources:
[a]Reference 1, p. 11.
[b]Reference 2, pp. 685-686.
[c]A. Sharp, *Bicycles and tricycles* (London: Longmans, Green & Company, 1896), p. 262.

Figure 1.6
Peak human power output for different durations.

A: Cycling (estimated) data from T. Nonweiler, "The work production of man: studies on racing cyclists," Proceedings of the Physiological Society, 11 January 1958, pp. 8P-9P.

B: Ergometer data from Loughborough University, personal communication (later data than reported on page 22).

C: Ergometer data from T. Nonweiler, "Air resistance of racing cyclists," The College of Aeronautics, Cranfield, England, report no. 106, October 1956.

D: Winch data from reference 2.

E: Ergometer hand-crank data from E. A. Müller, "Physiological methods of increasing human physiological work capacity," *Ergonomics,* vol. 8, no. 4, 1965, pp. 409-424.

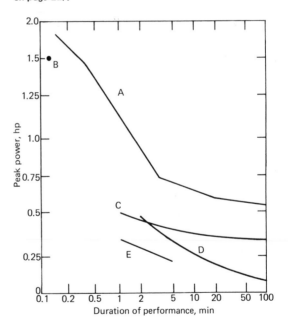

power output is associated with a moderate fraction of the maximum breathing capacity of the bicyclist. This power level can be shown by direct experiments to move a bicyclist and machine on the level at 9 to 13 mile/h [4.0 to 5.8 m/sec], depending on wind resistance and the condition of the road surface. This range of speeds has been long associated with average cycling since the standardization of good rear-driven pneumatic-tired bicycles.

Recently the breathing rates of pedaling bicyclists have been measured. Adams describes such experiments with riders moving at 10 mile/h [4.47 m/sec] and using 0.1 hp [74.6 watts].[3] Wyndham et al. show that at about this power output rather less than half the breathing capacity of an average man is involved,[4] and informed opinion now suggests that this exertion is the maximum which could be expected without adverse effects on health for average men working for long periods.

Information on the energy cost of locomotion of animals other than men can be found in references 5,6, and 7.

In a review of the energy used per ton-mile (or tonne-km) and passenger mile (km) for such varied means of transportation as the S. S. Queen Mary, the supersonic transport, a rapid-transit system, and oil pipelines, Richard Rice points out that a bicyle and rider are by far the most efficient.[8] He calculates that a modest effort by a bicyclist which results in 72 miles (116 km) being covered in 6 hours could require an expenditure of about 1800 calories (7,000 Btu). Assuming a weight of 200 pounds [90.6 km] for the rider and machine, Rice states that this figure is equivalent to over 1,000 passenger-miles [1,609 passenger-km] or 100 ton-miles [146 tonne-km] per gallon [3.785 liters] of equivalent fuel. The Queen Mary managed, by contrast, between 3 and 4 passenger-miles per gallon [1.27-1.70 passenger-km per liter].

**References
Chapter 1**

1. M. G. Bekker, *Theory of land locomotion* (Ann Arbor, Mich.: University of Michigan Press, 1952).

2. J. C. Trautwine, *The civil engineer's reference book,* 21st edition (Ithaca, N. Y.: Trautwine & Company, 1937), pp. 685-687.

3. W. C. Adams, "Influence of age, sex and body weight on the energy expenditure of bicycle riding," *Journal of Applied Physiology*, vol. 22, 1967, pp. 539-545.

4. C. H. Wyndham et al., "Inter-and intra-individual differences in energy expenditure and mechanical efficiency," *Ergonomics,* vol. 9, no. 1, 1966, pp. 17-29.

5. C. L. M. Kerkhoven, "Kenelly's law," *Work Study and Industrial Engineering,* vol. 16, February 1963, pp. 48-66.

6. Knut Schmidt-Nielson, "Locomotion: energy cost of swimming, flying and running," *Science,* vol. 17, 21 July 1972, pp. 222-228.

7. S. S. Wilson, "Bicycle technology," *Scientific American,* vol. 228, March 1973, pp. 81-91.

8. Richard A. Rice, "System energy and future transportation," *Technology Review,* Massachusetts Institute of Technology, Cambridge, Massachusetts, vol. 75, January 1972.

**Additional
recommended
reading**

Gans, Carl. "How snakes move," *Scientific American,* June 1970, pp. 82-96.

2

Human power generation

A great deal of experimental work has been carried out in scientific laboratories in order to assess the effort expended by humans in carrying out various tasks in different environmental conditions. Permissible work loads for the lunar astronauts were decided in this way. The impeding effect of the wearing of industrial protective clothing and the general state of health of both invalids and athletes can be assessed by giving the subjects a task to perform upon a machine to which is attached a power-measurement device—a machine called an "ergometer."

Ergometers

The measurement of power capacity exerted by a human via his legs is considered to be the most convenient means of assessing his physical work ability. The pedaling ergometer is a very satisfactory means of converting leg motions into measurable power.

Most ergometers have a frame, saddle, handlebars, and cranks similar to those of an ordinary bicycle. The cranks drive some form of resistance or brake, and the whole device is fastened to a stand which remains stationary during use (Figures 2.1, 2.2).

The methods employed for power measurement range from the crude to the sophisticated. One problem is that human leg-power output varies cyclically (as does that of a piston engine) rather than being smooth (as for a turbine). A device indicating instantaneous power (pedal force in the direction of motion multiplied by pedal velocity) would show peak values of perhaps a horsepower [746 watts] when the average is only 0.2 hp [149.2 watts]. Therefore some form of averaging is usually employed. In some cases the subject is supposed to keep his or her pedaling

Figure 2.1
Pedaling ergometer.
Courtesy of Renold Ltd.

Figure 2.2
Racing-bicycle ergometer.

rate constant over a minute or two to obtain accurate results. In other systems the power can be integrated and averaged electronically over any desired number of crank revolutions.[1,2]

A second problem with the measurement of human capability is the characteristic that, although human beings are extremely adaptable to an extreme range of power-transmission devices and conditions, they will train their muscles and responses to some degree of effectiveness with repeated use of any one device. If the power-measuring ergometer does not reproduce all of the conditions of bicycling, for example, it feels strange to the rider, and a sharply reduced output may be the result. Bicycle exercisers, which are sometimes fitted with "calorie meters" which make them crude ergometers, seldom reproduce the inertia of a bicycle rider and his machine, or the "viscous" or velocity-dependent drag, so that a pedaling technique completely different from the normal must be used, for instance, to get the cranks smoothly over "top dead center."

A third area of possible discrepancy between results obtained on ergometers and in practice is the effect of cooling of the subject.[3,4] A bicyclist is normally cooled by the relative wind he causes; when he pedals an ergometer, there may be almost no air movement, and his output may thereby be limited by heat stress. This topic is discussed more fully in Chapter 3.

Therefore, if tests by different experimenters of similar subjects show varied results, it seems justified to give more credence to the higher-power findings, assuming of course that the measurements have been accurately made.

Results from ergometer tests

Students at Dartmouth College in Hanover, New Hampshire, used an ergometer to find out what power output an ordinary nontrained bicyclist could maintain over useful periods of time.[5] It was found that for prolonged periods about 0.05 hp [37.3 watts] could be maintained with pedaling rates of 20 to 60 revolutions per minute (Figure

2.3a). It can be calculated that this power would give a bicycle road speed, on the level with no wind, of about 8 mile/h [3.6 m/sec]. This is a speed commonly achieved by an average "utility" bicyclist and therefore provides a check on the power measurement. This power result and other powers tolerable to the Dartmouth student bicyclists for briefer periods are shown as the nearly straight lines on Figure 2.3b. The expenditure of 0.05 hp [37.3 watts] can be achieved over a range of pedaling speeds from about 30 rpm to 60 rpm. Therefore a carefully selected "optimum"-performance gear is not necessary for the machine of a utility bicyclist, as indeed experience has shown. Japanese experimental work[6] confirming this finding is plotted on Figure 2.4.

Other ergometer experiments, similar in character to that associated with the report from Dartmouth College, are summarized in a table given in reference 10, p. 8. The subjects used by Dr. D. R. Wilkie of the University of London were instructed to exert themselves to their limit in order to record their maximum power outputs for varying periods of time. The peak power obtained was 0.54 hp [402.7 watts] for 1 minute and for 60 to 270 minutes the powers were 0.28 to 0.19 hp (208 to 141.7 watts).These powers are somewhat above those of the Dartmouth students and are similar to those recorded for laborers turning winches, as shown in the data recorded on Figure 1.6. It appears more logical, however, to take the Dartmouth College results as the more indicative of the power output of an average untrained utility rider.

Some tests with hand-powered ergometers (both hands cranking) are described by Müller.[8] A power output for a prolonged period of about 0.05 hp [37.3 watts] was recorded. (Compare Table 1.3).

Extensive experimental work with ergometers has been carried out at Loughborough University of Technology.[9] To date the vital importance of the use of a correct saddle height in relation to

Figure 2.3
Dartmouth College ergometer tests. The relationship of pedaling speed to torque is shown in part a. Power output as a function of pedaling speed is shown in part b. The horizontal straight lines in b show maximum power for an average pedaler for the durations noted. The curved lines are from data from reference 34.

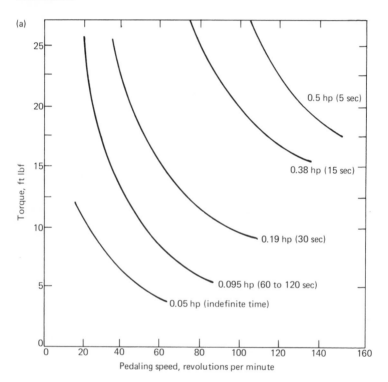

(a)

Torque, ft lbf

0.5 hp (5 sec)

0.38 hp (15 sec)

0.19 hp (30 sec)

0.095 hp (60 to 120 sec)

0.05 hp (indefinite time)

Pedaling speed, revolutions per minute

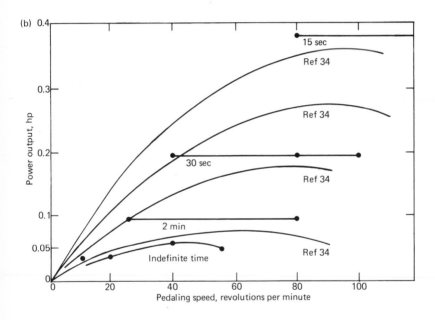

(b)

Figure 2.4
Power-output tests from
several sources.

Curve	Peak efficiency (%)
A	12.5
B	18
C	22
D	17
E	26
F	26
G	26
H	Optimum pedaling rates for the range of power outputs

Data for curves A, B, C, E, and F from reference 6.

Data for curve D from reference 32.

Curve G is estimated by FRW from data in reference 6.

Curve H is extrapolated from data in reference 6.

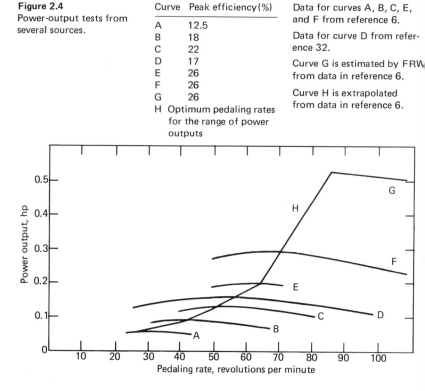

the pedals has been proved, quantitatively, by the test results. Already the work has given rise to the published figure of 1.5 hp [1119 watts] output for one bicyclist for a brief period of 5 seconds. This is recorded as the isolated point B on Figure 1.6 and is important in showing that the gap between the curves given in Figure 1.6 is likely to lessen as more ergometer experimental work is done, as was suggested above.

**Other methods of
power determination**

There are two other practical methods of assessing the power which a bicyclist can deliver. One is to measure the oxygen intake; this will be discussed later. The other way of estimating the power exerted by a bicyclist is to calculate it from the times achieved over measured distances combined with measured resistances of wheel rolling and air friction which have been obtained from separate experiments. Wind-tunnel investigations carried out at Cranfield College, England, recently have provided invaluable data on the wind resistance presented to racing bicyclists under various conditions.[10] Tire-rolling resistance has been measured by manufacturers over the years. It can be calculated that at 40 mile/h [17.88 m/sec] a racing bicyclist exerts about 1.5 hp [1119 watts] in overcoming wind and rolling resistances and mechanical friction. This can be verified by the fact that mopeds fitted with engines of about 1.5 bhp [1119 watts] output achieve about 35 mile/h [15.65 m/sec]. (See Table 1.1).

All information quoted previously may be correlated in a graphical summary of power output against forward speed. This is given as Figure 2.5 and is based on a rider of 150 lbm* [68 kg] with a machine of 20 lbm [9.07 kg] riding in a racing (crouched) position. If the rider adopted a more upright position and rode a more nearly "average" machine, it is probable that the powers required would be increased by about 20 percent on the whole. Such a rider could obviously not reach the speed of 40 mile/h [17.88 m/sec] shown, and confirmed in practice, to be possible

---

*We are using the notation "lbm" for "pounds mass," and "lbf" for "pounds force." Most systems of units have different units for mass and force, but in the English ("Imperial") system the common unit of "pound" occasionally causes confusion unless this differentiation is used. The two types of pound are connected through Newton's first law:

$F = ma/g_c$

Here F can be expressed in pounds force, and m in pounds mass. The acceleration a is in feet per second^2 and g_c is a constant equal, for Imperial units, to 32.174 lbm ft/lbf sec^2. A pound weight is defined as the force of attraction due to the earth's gravity on a pound mass at sea level (where a, the local value of the gravitational acceleration, is 32.174 ft/sec^2). The force we call weight at other levels of gravity (for instance at a high elevation on a mountain) is given by Newton's law by substituting the local level of gravitational acceleration for a.

In S. I. units, g_c is unity. Force (and weight) is exfor short periods with "racing" machines and racing positions.

Breathing processes—the human "engine"

The power from muscular action derives from the "burning up" of human-tissue chemicals via the oxygen taken in from the air in the lungs. A typical experimental investigation, as described by Wyndham et al., shows that an average man pedaling on an ergometer produces about 0.1 hp [74.6 watts] for every liter per minute of oxygen absorbed by his lungs from the air breathed (at a rate of about 24 liters of air per liter of oxygen) above that oxygen absorbed when he is at rest.[11]

Laboratory experiments on the calorific value of the chemicals known to be associated with the human breathing processes have shown that when one liter per minute of oxygen is used, a power output of about 0.4 hp [298.3 watts] should be obtainable. Muscular action is, therefore, said to be only about 25 percent efficient. The rest of the missing energy appears as heating effects in the human body, causing it to produce perspiration in order to keep its temperature down to a tolerable limit. The extent to which the body temperature can rise through heavy work is a controversial matter. A common view is that no more than $3.6°$ F $(2°$ C$)$[12, 13-17] is acceptable in most circumstances. It is desirable therefore for

---

pressed in newtons, and mass in kilograms. The acceleration is in meters per second^2.

Some additional confusion may arise in the distinction between mass and weight because we often loosely refer to the weight of a person or object when we really mean the mass. For instance, when we calculate the power output of a rider "weighing" 150 pounds or the power necessary to accelerate a 25-pound bicycle, it is actually the mass that we must use. The rider and the bicycle would both weigh less on the moon, but the rider's muscle output and the power needed for acceleration would not change on account of the reduced gravity. Hence most "weights" will be translated into pounds mass (lbm) and kilograms (kg). However, when the additional energy required to move a rider or bicycle up a hill is being calculated, it is appropriate to give the weight as pounds force (lbf) and newtons.

Figure 2.5
Walking and bicycling
power requirements.

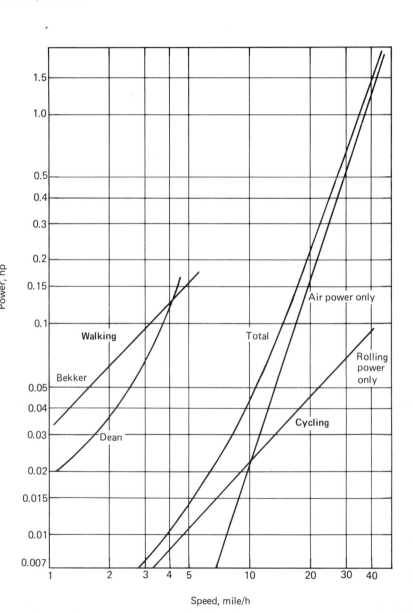

Speed, mile/h

exercising humans that their perspiration evaporates and does not just drip off their skin. Fast-moving air evaporates water far more quickly than slow-moving air. As a consequence a pedaler on a stationary ergometer can, and does, drip sweat profusely at a work rate of 0.5 hp [372.8 watts]. At 27 mile/h [12.1 m/sec], a speed corresponding to 0.5 hp (372.8 watts), an actual riding bicyclist is cooled far more effectively by sweat evaporation.[1,8]

Thermodynamic engines such as steam-turbine plant and internal-combustion engines are also usually only about 20-45 percent efficient in converting fuel energy to mechanical work. However, the limitations here derive from the second law of thermodynamics,* and therefore from the levels at which heat is added to and rejected from the engine. A steam turbine fed with high-temperature steam is more efficient than one using steam at lower temperatures. To achieve a thermodynamic efficiency of 25 percent, even an ideal engine, rejecting heat at room temperature T_1, as must the human body, would require that its fuel energy be absorbed at T_2, which can be calculated as follows:

---

*One of the many ways of expressing the second law of thermodynamics is the following. No engine can be more efficient than a thermodynamically reversible engine. And the efficiency of such an engine can be shown to be given by

$$\text{efficiency} \equiv \frac{\text{power output } w}{\text{rate of heat input } Q_2}$$

$$= \frac{Q_2 - Q_1}{Q_2} \text{ for all engines}$$

$$= \frac{T_2 - T_1}{T_2} \text{ for reversible, perfect engines,}$$

where T_2 is the temperature and Q_2 is the rate of heat addition, and T_1 the temperature and Q_1 the rate of heat rejection. Temperatures are given in degrees above absolute zero (degrees Rankine, $^\circ R$, on the fahrenheit scale [degrees Kelvin, $^\circ K$, on the Celsius (centigrade) scale]). Absolute zero is $-460\,^\circ F\,[-273\,^\circ C]$.

$$\text{efficiency} \equiv 1 - \frac{T_1}{T_2} = 0.25 = \frac{1}{4}$$

$$T_2 = \frac{4\,T_1}{3} = \frac{4 \cdot 540\,^\circ R}{3} = 720\,^\circ R = 260\,^\circ F$$
$$[127\,^\circ C]$$

for a room temperature T_1 of 80 °F (540 °R, 27 °C).

Obviously 260 °F cannot be tolerated in the body. Therefore the human "engine" is one that is not subject to the restrictions of the second law of thermodynamics. It is a type of fuel cell in which chemical energy is converted directly to mechanical power.

The energy not converted to power must appear, as for heat engines and fuel cells, as heat.

The human engine has, however, an additional characteristic not generally found in man-made machines in that some fuel must be "burned" to keep it going when it is at rest. (There is some similarity in a traditional steam plant, in which fuel must be burned continually to keep steam pressure up even when no power is being delivered.)

For a man of average weight the oxygen absorbed by his lungs when he is at rest and apparently not using any muscles is about one-third of a liter per minute. This quantity is additional to any other absorption consequent upon the man exercising his muscles.

In ordinary air, one liter of oxygen is found in a total of about five liters of air. However, when air is breathed, about twenty-four liters must be passed through the lungs for one liter of oxygen to be absorbed. (This average value of the "ventilatory equivalent" is given by Knipping and Moncreiff.[19]) Thus it is seen that about 380 percent more "excess" air than is needed to produce energy is used in the human engine. Most other engines, such as internal-combustion and steam engines, require only about 5 to 10 percent excess air to ensure complete combustion of the fuel.

Gas turbines more nearly approach human lungs in taking in about 200 percent excess air.

Cycling-speed versus breathing-rate relationships

Using all the information previously mentioned Table 2.2 has been drawn up showing how breathing rates increase for an average rider (150 lbm [68.04 kg]) cycling on the level in still air. It is assumed that for every liter of oxygen absorbed 24 liters of air have to be breathed.

For a nonathletic person the maximum oxygen breathing rate is assumed to be about 3 liters per minute. Table 2.2 shows that when a rider is using about half his maximum oxygen breathing capacity his power output is about 0.1 hp [74.6 watts]. These conditions are thought to be such that an average fit man could work for several hours without suffering fatigue to an extent from which reasonably rapid recovery is not possible. This rate of work is recommended for workers in mines.[20, 21] Experience has also shown that 0.1 hp [74.6 watts] propels a rider at about 12 mile/h [5.36 m/sec] when riding a lightweight touring machine (see Chapter 1). As this speed is one which can ordinarily be maintained by experienced but average touring-type riders, the numbers given in Table 2.2 seem sound. (See also miscellaneous data given by Adams[22] and Harrison et al.[23] which show average heat loads of 4.2 - 9 kg cal/min for speeds 6½ - 13 mile/h [2.9 - 5.8 m/sec]. Some of these and other data have been collected in Figure 2.6.

Maximum performance

The performance of athletes has been investigated, and maximum oxygen intakes of much greater than 3 liters per minute have been recorded. In addition it appears that training can accustom a fit man to work at greater than 50 percent of his maximum intake for prolonged periods.

The potential power output of humans (and all other animals) is time dependent (Figure 1.6). Physiological experimenters credit some aspects of this phenomenon to the body's capability to call upon a type of reserve oxygen supply in

Figure 2.6
Gross caloric expenditure by bicyclists.
Data for points from—
L. Zuntz, *Untersuchungen über den Gassweschel und Engegeseumsatz des Radfahrer* (Berlin: Hirschwald, 1899);
D. B. Dill, J. C. Seed, and Z. N. Marzulli, "Energy expenditure in bicycle riding," *Journal of Applied Physiology,* vol. 7, 1954, pp. 320-324;
O. G. Edholm, J. G. Fletcher, E. M. Widdowson, and R. A. MacCance, "Energy expenditure and food intake of individual men," *British Journal of Nutrition,* vol. 9, 1955, pp. 286-300;
M. S. Malhotra, S. S. Ramaswany, and S. N. Ray, "Influence of body weight on energy expenditure," *Journal of Applied Physiology,* vol. 12, 1962, pp. 193-235;
W. C. Adams, reference 21. Curves A and B from reference 3.
Curve C is based on data from reference 31 and J. S. Haldane, *Respiration* (London: Oxford University Press, 1922).

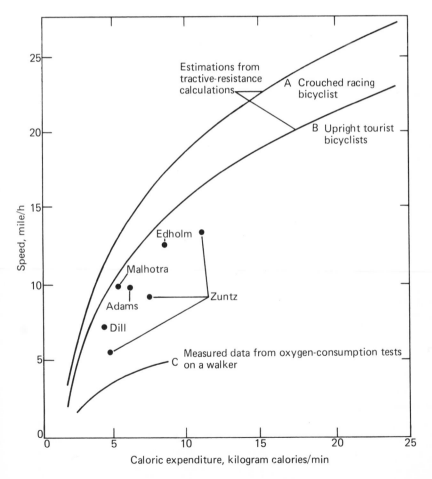

addition to that taken in by the lungs. When someone is drawing upon this reserve, for instance at the start of a 100-meter dash, he is said to be using oxygen in an "anaerobic" manner. That part of his oxygen supply which comes from breathing is "aerobic." Most of the energy given out after a minute is derived from aerobic breathing. Other physiological phenomena decide the tail-off of power output over prolonged periods of an hour upwards.

Experimental investigations and reports seem to have been confined to the study of a maximum period of exercise of about 100 minutes. Wilkie[24] and others quote no longer periods. However, useful estimates for longer periods can be made by analyzing British time-trial bicyclists' performances for rides of up to 24 hours duration.

Every year, in time-trial competitions, speeds of about 23 miles per hour [10.3 m/sec] are achieved by a few riders for 12-hour events (riding unpaced) and for 24 hours several achieve nearly 20 miles per hour [8.94 m/sec]. Using data from figures and tables in this chapter it can be estimated that these riders are working at 0.33 and 0.22 hp [246 and 164 watts]. These two outputs fit in well with the predicted trend proposed by Wilkie[25] and with extrapolations of the curves of Figure 1.6. Skiers have worked at an 80 percent level for hours on end.[26] Bicyclists riding at 25 mile/h [11.18 m/sec] for long periods are likewise using about the same proportion of their maximum oxygen intake.

Maximum oxygen intake is affected by age to the extent of being about half at 80 years compared with that at the peak of about 20 years old.[27]

Muscle efficiency and the effect of different movements

From information given particularly in references 28 and 29, but also supported by that in references 30, 31, and 32, it appears that whatever the muscular movement—pedaling, stepping up and down, or walking up gradients—the usage of oxygen for a given power output is similar. At power outputs of 0.1 to 0.3 hp [74.6 - 223.7 watts], the maximum output experimented upon, the net muscular efficiencies are all about 27 percent. This efficiency should be about that of touring bicyclists.

In view of the above it appears that the pedaling action does not suffer from any particular loss of power through back-and-forth leg swinging, as has often been suggested by advocates of up-and-down lever-propulsion mechanisms. The motion of the latter can be likened to the stepping movements experimented upon and described by McDonald.[33]

Optimum pedaling rates

The Dartmouth College results for higher power outputs, tolerable to untrained riders for brief periods only, are shown as almost straight lines on Figure 2.3b. The pedal push of the average rider is limited to about his own weight and therefore high power must be obtained by high pedaling rates. However, each individual has a certain maximum pedaling rate, even with zero resistance. Hence the limits of pedaling speeds given by the lines on the graph.

Brown propounded the existence of peaks in the power-output pedaling-rate curves of bicylists, although no description is given of the experimental origin of the data.[34] These peaks can be seen in the curves labeled "reference 34" on Figure 2.3b. It appears that the Dartmouth College work gives some support to the existence of limited ranges of pedaling rates for which certain power outputs can be obtained, though not pinpointing the peaks.

Information given by Garry and Wishart[35] in connection with the measured breathing and oxygen-absorption rates of riders of ergometers at power outputs of about 0.1 hp [74.6 watts], which is in the range common for nonracing bicyclists, gives interesting conclusions. It appears that a maximum muscle efficiency is achieved by pedaling at about 50 revolutions per minute. However, there is only a small drop in efficiency with a pedaling rate at about 30 percent either side of this maximum, that is, in the range 33 to 70 revolutions per minute (see Figure 2.4 curve D).

In contrast to the above findings, however, it is accepted that when power outputs of about 0.2 hp [149.1 watts] and beyond are required the pedaler must increase his pedaling speed above 50 revolutions per minute. It appears that there is a limit to the foot thrust that a pedaler can apply over prolonged periods; hence pedaling speed must go up to get increased power. According to Table 2.1 the tolerable thrust is about the equivalent to a tangential force of 54 lbf [240.2 newtons] for riding periods of about one hour. The average actual vertically applied thrust could be up to about double this value (108 lbf) through poor skill in pedaling involving wasteful thrusting at the dead centers, which just lifts the bicyclist's body.[36]

Limited data in connection with the short-distance high-climbing performances of bicyclists for some short periods of about 4 to 10 minutes show that power outputs of about 0.7 hp [522 watts] are common with road speeds (on 60- to 70-in. [4.8 - 5.6 m] gears) of about 12 miles per hour [5.36 m/sec]. In these cases the pedal thrusts are considerable, peaking at well over body weight, achieved, no doubt, by the rider finding it feasible to pull hard against his handlebars for the brief times involved.

Pedaling forces

Table 2.1, which was compiled from data given in other parts of this book, compares the recorded pedaling rates of bicyclists of all types with estimates of the power outputs. These estimates in turn have led to estimates of the tangential forces at the pedals resisting the motion.

It appears that paced bicyclists tend to use very consistent but moderate pedal thrusts amounting to mean applied tangential forces of only about one-fifth of the rider's weight. The peak vertical thrusts are greater[37] but are still relatively small. No doubt this action results in the rider being able to maintain a steady seat and an ability to steer steadily. This is vital at speeds of 50 to 100 miles per hour [22.3 to 44.7 m/sec].

Table 2.1. Pedaling speeds.

Cycle	Distance, miles	Time, min sec	Subject	Speed, mile/h	Gear, in.	Crank, in.	Crank speed, rpm	Foot speed, ft/min	Est'd power, hp	Est'd thrust, lbf
Ordinary, track	1/4	30 sec	R. Temple	30	53	5	190	493	1.35	91
	1/2	72 sec	U. L. Lambley	25	56	5	150	392	1.05	88
		60 min	H. L. Cortis	20.1	59	5 1/2	116	330	0.5	50
Safety, track	1/8	12.4 sec	H. Ryan	36.3	90	6 1/4	136	446	1.6	120
	1/8	12.2 sec	A. A. Zimmerman	37	68	6 1/2	182	619	1.6	85
	1/4	29 sec	A. A. Zimmerman	29.8	64	6 1/4	170	520	1.3	83
	1/8	11.5 sec		39	90	6 1/2	145	473	1.65	115
Safety, track		60 min	T. Verschueren	40.1	106	6 3/4	126	445	0.5	37
		60 min	H. Grant	56	139	6 1/2	134	456	0.5	36
motorcycle		60 min	A. E. Wills	61.5	144	6 1/2	143	488	0.5	35
paced		60 min	L. Vanderstuyft	71	180	6 1/2	133	454	0.5	36
		60 min	L. Vanderstuyft	76	191	6 1/2	134	454	0.5	36
Train-paced	1	57 sec	C. M. Murphy	62	104	6 1/2	198	670	1.2	59
Road safety bicycle	25	52 min		28.8	90	6 7/8	102	370	0.6	54
	100	4 h		25	85	6 7/8	99	368	0.5	45
	480	24 h		20	80	6 7/8	84	310	0.25	26
	100	4 h 28 min	F. W. Southall	22.4	81	6 1/2	93	316	0.5	52
Road, tourist				10	68	6 7/8	50	180	0.09	16
				12	68	6 7/8	61	220	0.11	16
				16	75	6 7/8	74	266	0.2	24
				18.5	75	6 7/8	85	305	0.3	32

Sources:

A. C. Davison, "Pedaling speeds," *Cycling*, 20 January 1933, pp. 55-56.

H. H. England, "I call on America's largest cycle maker," *Cycling*, 25 April 1957, pp. 326-327.

"'Vandy,' the unbeaten king," *Cycling*, 11 March 1964, p. 8.

Marcel De Leener, "Theo's hour record," *Cycling*, 7 March 1970, p. 28.

It is easy to calculate from the crank length and pedaling speed in revolutions per minute how much thrust upon the pedals is required for a given horsepower output. The peripheral pedal speed around the pedaling circle (or the vertical speed on the downward stroke) can be used in the equation

Thrust force (lbf)
$$= \frac{\text{power (hp)} \times 33{,}000 \text{ (ft lbf/min)/hp}}{\text{pedaling speed (ft/min)}} .$$

In S.I. units

$$\text{Thrust force (newtons)} = \frac{\text{power (watts)}}{\text{pedaling speed (m/sec)}} .$$

The senior author[38] has carried out ergometer experiments under constant-speed pedaling conditions in order to check the agreement between the measured thrust and the calculated thrust. It was found that at the optimum pedaling speeds (related to power outputs as in Figure 2.4) the measured thrust agreed with the predicted thrust to a reasonable accuracy, particularly for power outputs above 0.1 hp [74.6 watts]. At pedaling rates other than the optimum the measured average vertical thrust upon the pedal over its path was greater than that expected by amounts which could be predicted from the lowering in pedaling efficiency as given on Figure 2.4 by oxygen-consumption tests. Hence it was concluded that, at other than optimum pedaling rates, thrust is "wasted" somewhere in the system. Maybe the body is lifted unnecessarily or the legs are swung so that lateral thrust components occur. Hoes et al. have found that at 60 revolutions per minute, measured pedal thrusts are near those expected from the ergometer power requirements.[39]

It is known that pedaling does involve foot thrust components other than the simple vertical ones, but apparently these are relatively small in effect and, at least at power outputs of 0.15 hp [119 watts] onwards, the simple vertical thrust predominates.

Measurements made during actual bicycling

Thorough and accurate data relating oxygen consumption, heart rate, pedal torque, pedaling rate, bicycling speed, gear ratio, and crank length have been taken by the Japanese Bicycle Research Association by equipping several riders with instruments and recording their behavior during actual riding by means of telemetry equipment.[40] The results tend generally to substantiate the foregoing discussion. Figure 2.7 shows the relation between oxygen consumption and heart rate for four subjects ranging from a trained athlete to an everyday working bicyclist. The best performance of one of the competition bicyclists over various distances using a range of gear ratios is shown in Figure 2.8. The best times and speeds were given when the highest gear ratio was used—about 111 inches [8.85 m] —except for the shortest distance, 200 meters, for which a range of gear ratios gave virtually identical average speeds. Tests of different crank lengths were inconclusive, but tended to show best performances with a crank length of 6¾ in. (approximately 170 mm*) for the untrained bicyclists, and to show no significant effect of crank length on average speed over 1000 meters for the trained riders.

An interesting cross correlation of efficiency versus crank speed for various average speeds and gear ratios is shown in Figure 2.9 for the bicyclist who produced the most work per liter of oxygen consumption. The peak efficiency (about 30 percent) at the higher speeds (30 and 35 km/h) [8.33 and 9.72 m/sec] was obtained at 60 to 70 crank rpm and at the highest gear ratio of 111 in. [8.85 m].

A tentative conclusion is that racing bicyclists could use gear ratios higher than those usually employed, since peak efficiency was not reached even at 111 inches [8.76 m]. Such gear sizes are coming into use as the top gears of multispeed

---

*While cranks of 6¾ in. are closer to 171 mm long, we have given the nearest crank available in metric sizes, 170 mm.

bicycles nowadays, particularly for time-trial racing. The Japanese riders did, however, complain of leg strain and it may be advisable for most riders to continue with their time-proved slightly lower gears which give pedaling rates of some 90 to 100 revolutions per minute when ridden to the capacity of the individual.

Adams[41] and Whitt[42] give summaries of results obtained by earlier workers than the Japanese on actual measurements of oxygen breathing of bicyclists when in motion. This work of course involves a small handicap through the bicyclist having to carry equipment for the test; the speeds used on the road were but moderate. The Japanese telemetry equipment was very light and probably brought about insignificant changes in performances.

Cycling versus roller skating

From Figure 2.10 it can be seen that for equal periods of maximum power output the record speed credited to a roller skater is less than that of a track bicyclist, being, for two minutes effort, 21 mile/h [9.39 m/sec] compared with 35 mile/h [15.65 m/sec]. If it is assumed that such record

Figure 2.7
Oxygen-consumption measurements during bicycling. Data from reference 6.

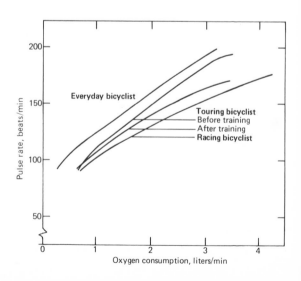

Figure 2.8
Effect of bicycle gear ratio
on performance. From
reference 6.

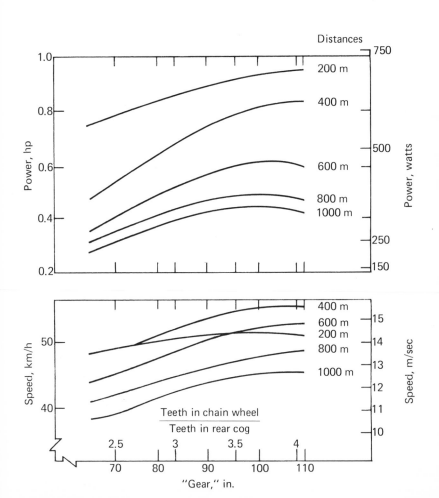

makers exert equal powers at their respective relative speeds, an estimate can be made of the rolling resistance of roller skates as follows.

Assume that the skater has a frontal area of 3 sq ft [0.279 sq m], which is less than the 3.65 sq ft [0.339 sq m] of a very crouched bicyclist and machine. At 21 mile/h [9.39 m/sec] a bicyclist exerts 0.2 hp to overcome air resistance (see Figure 2.5). Therefore

power needed by the skater to overcome air resistance

$$= \frac{3}{3.65} \times 0.2 \text{ hp}$$

$$= 0.164 \text{ hp } [122.3 \text{ watts}].$$

Figure 2.9
Effect of gearing on energy efficiency. From reference 6.

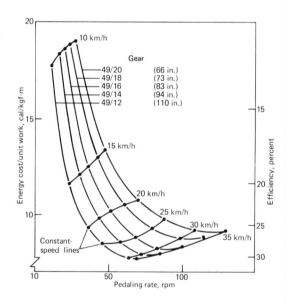

At 35 mile/h [15.65 m/sec], the bicyclist exerts
1.1 hp [820 watts] (see Figure 2.10), and we
assume that the skater at 21 mile/h [9.39 m/sec]
is exerting the same power. Hence,

power absorbed by the skates

= 1.1 hp − 0.164 hp

= 0.936 hp [698.0 watts].

If the skater weighs 154 lbm [0.0687 long tons,
or 69.85 kg],

rolling resistance of skates

$$= \frac{0.936 \text{ hp} \times 375 \text{ (mile lbf/h)/hp}}{0.0687 \text{ long ton} \times 21 \text{ mile/h}} = 243 \text{ lbf/long ton.}$$
[1.0595 newtons
per kg]

Figure 2.10
World-record speeds by
human power in various
modes.

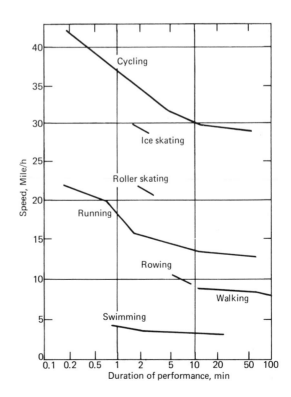

The above rolling resistance is very high compared with that of bicycle wheels, assumed for the purposes of the construction of Figure 2.5 as 11.5 lbf per long ton (0.0503 newtons per kg). The twentyfold increase can be partly explained as the effect of the use of very small wheels in the skates—about a thirteenfold decrease in diameter—and the rest to the less-easy running at high speed of the hard rollers compared with the pneumatic tires of the bicycle. Measurements of the pull required to keep a skater running steadily made by the senior author showed a rolling resistance value of about 134 lbf/long ton at low speeds, and other information suggests that this would increase greatly at speeds of 21 mile/h [9.39 m/sec].

There are several current attempts at producing skates having large wheels of much lower rolling resistance to determine the effectiveness of this form of man-powered locomotion. Cross-country skiers train in summer on a form of large-wheeled roller skate (see Figure 10.8).

Cycling versus walking (on the level, in still air)

For the purpose of comparison, Tables 2.2 and 2.3 have been drawn up from information given in references 43, 44, and 45 and elsewhere. The data of Dean,[46] which is the main source for Table 2.3, can be interpreted as meaning that the maximum tractive resistance of the walker is about 1/13th his weight. This figure was given as early as 1869.[47] A higher resistance of 1/7.5th is however estimated from a simple geometrical model.[48]

The tables show that for the same breathing rate the bicyclist's speed is about four times that of the walker.

The metabolic-heat figures were obtained by multiplying the oxygen consumption, in liters per minute, by a calorific-value constant of 5 kilocalories per liter of oxygen, given by Falls as a reasonable value for the circumstances.[49] This represents the total "burn up" of human tissue which must ultimately be replaced by food. If

Table 2.2. Cycling—breathing rates.

Speed, mile/h		Tractive power, hp	Breathing rate, liters/min		Metabolic heat, kcal/min
Racer	Tourist		Oxygen	Air	
27	22.5	0.5	4.8	115	24
25	21	0.4	3.4	93	19.5
22	18.5	0.3	3	72	15
19	16	0.2	2.1	50	10.5
14.5	12	0.11	1.2	29	6
10.5	8.3	0.05	0.75	18	3.75
7.2	6	0.025	0.53	13	2.65
3.2	1.8	0.008	0.38	9	1.9
0	0	0	0.3	7	1.5

Table 2.3. Walking—breathing rates.

Speed, mile/h	Tractive power, hp	Breathing rate, liters/min		Metabolic heat, kcal/min
		Oxygen	Air	
4.46	0.141	1.83	44	9.1
3.33	0.076	1.1	26	5.5
2.23	0.0415	0.71	18	3.5
1.1	0.0226	0.52	12.5	2.5
0	0	0.28	6.8	1.4

Sources for Tables 2.2 and 2.3:
References 43, 44, and 45.

each kilocalorie could be converted at 100 percent efficiency to mechanical energy (via muscular action), 0.09 hp [67.1 watts] should result.

Dean shows that walking up a hill is slightly more efficient an action from the point of view of oxygen consumption than is level walking,[50] so that the difference between cycling and walking is lessened in these circumstances.

Cycling versus running (on the level, in still air)

The recorded times for sprint runners and racing bicyclists on the track show that a bicyclist can reach 40 mile/h [17.88 m/sec] for the furlong (220 yards [201.17 m]) and 30 mile/h [13.41 m/sec] for the mile (1609.3 m) while a runner reaches only half these speeds. Assuming that the wind resistance of a bicyclist on his machine and of a runner are similar at similar speeds, we can estimate that the powers needed for cycling are only about one-fifth that the powers needed for running at the same speed (15 to 20 mile/h [6.7 to 8.9 m/sec]) range.

Gradient resistance

Gradients and headwinds impede both bicyclist and runner or walker but to different relative degrees compared with level progression in still air. It can be calculated that a gradient of 4 percent (1 in 25) or a headwind of 10 mile/h [4.47 m/sec] slows down a bicyclist exerting a constant 0.05 hp [37.3 watts] to about his slowest balancing speed of 2½ mile/h [1.12 m/sec]. A walker developing the same power would be slowed from about 2 mile/h [0.89 m/sec] to 1¼ mile/h [0.56 m/sec]. The rider is slowed down to 25 percent speed and the walker is slowed down to about 55 percent speed. As a consequence the bicyclist notices difficult conditions more vividly than does a walker. On the other hand, with tailwinds or downhill, the bicyclist is aided to a far greater extent than the walker, and it is probably this virtue of the bicycle that will ensure its use even as an aid to walking in very hilly country.

When a bicyclist or walker climbs a hill, his
weight has to be lifted through a vertical distance,
and as a consequence extra power is required
above that for progress along the surface of the
road. The additional power required from a bicy-
list with a total weight of 170 lbf [756 newtons]
to climb a hill of, say 5 percent (1 in 20) and at
25 mile/h [11.18 m/sec] is

$$\frac{170 \text{ lbf} \times 25 \text{ mile/h}}{20 \times 375 \text{ (mile lbf/h)/hp}} = 0.57 \text{ hp [425 watts]}.$$

Hence, it is seen from Figure 2.5 that a racing
bicyclist climbing a hill of 1 in 20 (5 percent)
must give out a power of 0.57 +0.407 or 0.97 hp
[723.3 watts]. He would be sorely stressed and
could do this for only about two minutes, according to Figure 1.6.

Bradley gives interesting information about
his climbing feat in the Tour of Austria.[51] Bradley
climbed a 1-in-12 (8.5 percent) pass on the Gross
Glockner of 12½ miles [20,117 m] length in
about 57 minutes. The gear used was 47 in. [3.76 m],
and it can be deduced that he exerted at least
0.6 hp [447.6 watts], pedaling at a rate of about
90 revolutions per minute. This performance is
remarkably close to the fast 25-mile [40,233.5 m]
time-trial performances shown in Table 2.4 and
provides most convincing proof that there is
sound evidence for all the power-requirement
estimates based on wind-resistance calculations,
as distinct from the more easily accepted simple
weight-raising calculations associated with hill-
climbing bicyclists.

**Should one walk or
pedal up hills?**

Noncompetitive bicyclists have the option of
walking up steep hills. Some prefer to do so,
alleging that a change of muscle action is agree-
able to them. Some bicyclists, however, prefer
to fit low gears to their bicycles and to ride as
much as possible.

Whether it is easier to ride or to walk up
steep gradients is a subject often debated

Table 2.4. Principal bicycling speed and time-trial records.

World's track records

Professional unpaced standing start:
1 km, Milan, 1952, R.H. Harris, 1 min 8.6 sec
1 h, Mexico, 1972, E. Merckz, 49.408 km

Amateur unpaced standing start:
1 km, Mexico, 1967, G. Sartori, 1 min 4.6 sec
1 h, Mexico, 1969, T. Radames, 46.95 km

Unofficial and unrestricted:
1 h, motor paced, standing start, Montlhéry, 1928, L. Vanderstuyft,
76 miles 503 yards
1 km, motor paced, flying start, Freiburg, 1962, J. Meiffret,
127.25 mile/h (204.77 km/h)

British amateur unpaced road records

Time trial competition		h	min	sec
Men				
Bicycle				
25 miles	A. R. Engers, 1969	0	51	0
100 miles	A. Taylor, 1969	3	46	37
12 h	E. J. Watson, 1969	281.87 miles		
24 h	R. Cromack, 1969	507.00 miles		
Tricycle				
25 miles	D. R. Crook, 1966	0	59	58
100 miles	E. Tremaine, 1971	4	30	48
12 h	H. Bayley, 1966	249.65 miles		
24 h	J. F. Arnold, 1953	457.33 miles		
Women				
Bicycle				
10 miles	B. Burton, 1967	0	22	43
100 miles	B. Burton, 1968	3	55	5
12 h	B. Burton, 1967	277.25 miles		
24 h	C. Moody, 1969	427.86 miles		
Tricycle				
10 miles	C. Masterson, 1961	0	28	51
100 miles	C. Masterson, 1958	5	17	9
12 h	J. Blow, 1960	212.82 miles		
24 h	J. Blow	374.15 miles		
Tandem				
30 miles	G. M. Tiley & J. Budd, 1951	1	11	36

Long distance		days	h	min
One thousand miles				
Bicycle	R. F. Randall, 1960	2	10	40
Tricycle	A. Crimes, 1958	2	21	37
Tandem bicycle	P. M. Swinden & W. J. Withers, 1964	2	18	9
Tandem tricycle	A. Crimes & J. F. Arnold, 1954	2	13	59
Land's End to John o' Groats (872 miles)				
High ordinary	G. P. Mills (P), 1886	5	1	45
Bicycle	R. F. Poole, 1965	1	23	46
Tandem bicycle	P. M. Swinden & W. J. Withers, 1966	2	2	14
Tricycle	D. P. Duffield, 1960	2	10	58
Tandem tricycle	A. Crimes & J. F. Arnold, 1954	2	4	26

Source: Cyclist Touring Club

among bicyclists. We will use the data previously
developed to show that it should be more efficient to ride up to an approximately limiting
gradient.

If we confine attention to the everyday bicy-
list, we can assume that he is unlikely to wish to
use much more than about 0.1 hp [74.6 watts].
A commonly encountered steep hill is one with
a gradient of 15 percent or 1 in 6.7. It is assumed
that the road speed which is thereby fixed as
1½ mile/h [0.67 m/sec] is one which gives no
difficulties from the balancing point of view.

**Overall efficiency of the muscular action of the
riding bicyclist:** Many experiments have been
carried out on the oxygen consumption of pedalers.[52,53] The data given in Figure 2.4 appear
typical in that for a power output of 0.1 hp [74.6
watts] at the wheel at metabolic (gross) efficiency
of 21 percent is reasonable. The bicyclist will be
"lifting" a machine weighing, say, 30 lbf [130
newton] in addition to his body (150 lbf [667
newton]), so that a factor is necessary for the
efficiency when compared to body weight alone.
This can be calculated as 21 × 150 lbf / (150 + 30)
lbf or 17.5 percent (assuming there is negligible
rolling or wind resistance at 1½ mile/h [0.67 m/sec])
Power losses in the low-gear mechanism are also
neglected (see later).

**Overall efficiency of muscular action of a walker
pushing the machine:** Macdonald gives a summary
of experimental work concerning the oxygen
consumption of walkers going up various gradients
at different speeds.[54] For a walking rate of 1½
mile/h [0.67 m/sec] up a gradient of 15 percent
it appears that a metabolic gross efficiency of 15
percent is accepted as typical. This efficiency
assumes as a basis the body weight being lifted
against gravity. The bicyclist pushing his machine
will be in a semicrouched position so that an
adjustment to the efficiency must be made. Data
from Dean[55] as well as from Macdonald[56] con-

cerning the effects of walking in stooping positions and when carrying small weights show that the pushing of the 30-lbf [130 newton] bicycle absorbs 30 percent extra effort so that the walker's muscle efficiency based on his body weight alone is decreased to $17.5 \times (100 - 30) / 100 = 12.3$ percent. From the estimations given above it appears that it is easier to ride up a 15 percent gradient than to walk at the same speed of 1½ miles per hour [0.67 m/sec] pushing the bicycle by about 12.3/17.5 or a reduction of 30 percent.

However, in practice the lowest gear available may be of 20 in. [1.6 m], giving a pedaling rate of 26 revolutions per minute, which according to figures 2.3 and 2.9 is not optimum. A lowering of the previously assumed 21 percent pedaling overall efficiency is bound to occur. Let us estimate this at about 18 percent. As a consequence the 30 percent difference previously quoted should be taken as about 18 percent. It must be realized that this difference gives but a small margin for the extra friction involved in the bicycle transmission using a very low gear. Calculations on the lines of the above show that the 15 percent gradient is perhaps a critical one and that at gradients of 20 percent, or 1 in 5, there is no really appreciable advantage in riding the cycle even though a low gear is used.

A matter not given prominence in this type of discussion is one concerned with the lack of wind cooling for a bicyclist's relatively high heat output. At a power output of 0.11 hp [82.0 watts] he would, on the level, be traveling at some 14 mile/h [6.2 m/sec] with the cooling experience of such a wind for level running. Under such circumstances he would be fairly comfortable and would not feel overheated. When climbing a hill at 1½ mile/h [0.67 m/sec] for a period of, say, a quarter of an hour, it is certain that an averagely clothed bicyclist will feel himself getting hot. Unpublished data suggest a body temperature rise of an appreciable magnitude—1 °F [0.55 °C]. It is probable that such considerations influence bicyclists to get off and walk at very low speeds of, say, less than 1 mile/h [0.45 m/sec] when the lower heat loss from the lowered power output is more tolerable.

Comparison of human power with internal-combustion engines and electric motors

Only two types of small power units have been developed for propelling light bicycles. The small internal-combustion engine of the "moped" is well known. The other type is a small electric motor supplied by lead-acid storage batteries. A recently marketed scooter for use in factories uses this system. A Humber pacing tandem was in use in the 1890s and was fitted with an electric motor and a frame full of batteries (Figure 2.11).

The specification for modern "mopeds" shows that a gasoline engine and accessories of power and endurance equivalent to a man would weigh about 20 lbm [9.07 kg]. Performance details for electric propulsion show that this method would add about the weight of a man in the form of batteries and small electric motor (with an output of about 0.1 hp [74.6 watts]).

S. W. Gouse gives a chart (Figure 2.12) of general data concerning specific powers and specific energies for various power sources.[57] We have added data for a racing bicyclist riding at 20 mile/h for 24 h; 22 mile/h for 12 h; and 25 mile/h for 4 h. A touring bicyclist covering about 100 miles in 8 hours is included as point •. There is a peculiar similarity between human energy capacity and that of lead-acid batteries, even though bicycling performances are not strictly comparable to those for batteries because the batteries are not recharged, as is the long-distance bicyclist in his periodic snacks.

Other data on various heat engines and human performances are given in Table 2.5.

General comments

This chapter has dealt with the "human machine" as distinct from the bicycle itself. From time immemorial men have investigated the power output of humans and animals. Farey summarized

Figure 2.11
Humber electric bicycle, 1898. Reproduced from *Motorcycle Story* by Harold Connolly with permission from Motor Cycle News Ltd.

Figure 2.12
Bicyclists compared with engines and batteries. Data from reference 57.

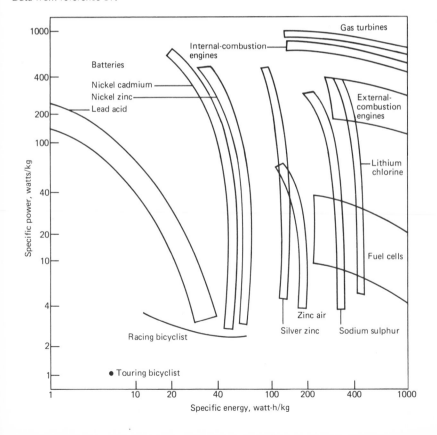

Table 2.5. Performance details of heat engines.

Engine type	Fuel	Fuel cons, lbm/bhp h min	Air consumption gross & net				Theoretical air-lbm/ lbm fuel	Thermal efficiency percent	Calorific value of fuel		Exhaust composition, dry		
			Throughput		Absorbed cu ft/hph	Oxygen absorbed liter/min. 0.1 hp							N2 and other gases
			liter/0.1 hp min	Cu ft/ hph					Btu/lbm	kg cal/g	CO2	O2	
Steam engine (non con- den- sing)	Coal												
	Coal	3.75	40	800	525	4.6	11.5	4.56	13,000	7.2	12.5	7.2	80.3
Steam engine (con- den- sing)	Coal	1.5	15	323	213	2.0		11.6					
Spark ignition (car)	Gaso- line	0.7	6	130	130	1.32	15	18.3	20,000	11.1	15	–	85
(aero)	Gaso- line	0.45	3.9	84	84	0.79		29					

Table 2.5. Performance details of heat engines (continued).

Engine Type	Fuel	Fuel cons, lbm/bhp h min	Air consumption gross & net				Theoretical air-lbm/lbm fuel	Thermal efficiency percent	Calorific value of fuel		Exhaust composition, dry		
			Throughput		Absorbed cu ft/hph	Oxygen absorbed liter/min 0.1 hp			Btu/lbm	kg cal/g	CO_2	O_2	N_2 and other gases
			liter/0.1 hp min	Cu ft/ hph									
Diesel marine (large)	Oil	0.33	3.8	80	60	0.56	14.6	39.5	19,000	10.5	10.6	5.5	83.9
Human	Fat Carbo-hydrate	1.04	26	540	116	1.1	9.0	21	12,000	6.75	4.5	16	79.5

Sources:

Kempe's engineers year book, vol. II (London: Morgan Brothers, 1962), pp. 153, 380, 384, 389, 575, 584, 596.

R. S. McLaren, Mechanical engineering for beginners and others (London: Charles Griffin & Company, 1917), pp. 276, 401-402.

C. H. Best and N. B. Taylor, The physiological basis for medical practice (London: Baillieres, Tindall & Cox, 1939). p. 849.

Mechanical world year book, 1967 (Manchester: Emmott & Company, 1967), p. 158.

Sir Richard Glazebrook, editor, A dictionary of applied physics, vol. 1 (London: Macmillan & Company, 1922), p. 689.

J. S. Haldane, Respiration (London: Oxford University Press, 1922), p. 156.

C. H. Best and N. B. Taylor, The physiological basis for medical practice (London: Baillieres, Tindall & Cox, 1939). p. 849.

such data, obtained prior to the 19th century, recognized to be of value to engineers.[58] Since then specialized workers have obtained more refined experimental data and a very large amount of recorded data are available. Reference 59 includes textbooks summarizing such experimental findings concerned with healthy humans. However, these textbooks do not concentrate on pedaling bicycles. To assemble data on bicycles, much sifting of relevant generalized literature must be carried out. Vaughan Thomas, discussing pedaling rates,[60] is one of the few to admit that practicing bicyclists are conservative in their views about accepting "everthing that the boffins (scientists) tell them."

**References
Chapter 2**

1. Wilhelm von Döbeln, "A simple bicycle ergometer," *Journal of Applied Physiology,* vol. 7, 1954, pp. 222-224.

2. C. Lanooy and F. H. Bonjer, "A hyperbolic ergometer for cycling and cranking," *Journal of Applied Physiology,* vol. 9, 1956, pp. 499-500.

3. F. R. Whitt, "A note on the estimation of the energy expenditure of sporting cyclists," *Ergonomics,* vol. 14, no. 3, 1971, pp. 419-424.

4. D. Clifford, D. McKerslake, and J. L. Weddell, "The effect of wind speed on the maximum evaporation capacity in man," *Journal of Physiology,* vol. 147, 1959, pp. 253-259.

5. "Report on the energy-storage bicycle," Thayer School of Engineering, Dartmouth College, Hanover, New Hampshire, 1962.

6. "Report of the Bicycle Production and Technical Institute," Japan, 1968.

7. D. R. Wilkie, "Man as an aero-engine," *Journal of the Royal Aeronautical Society,* vol. 64, 1960, pp. 477-481.

8. E. A. Müller, "Physiological methods of increasing human work capacity," *Ergonomics,* vol. 8, no. 4, 1965, pp. 409-424.

9. Loughborough University, private communication.

10. T. Nonweiler, "Air resistance of racing cyclists," The College of Aeronautics, Cranfield, England, report no. 106, October 1956.

11. C. H. Wyndham et al., "Inter-and intra-individual differences in energy expenditure and mechanical efficiency," *Ergonomics,* vol. 9, no. 1, 1966, pp. 17-29.

12. See reference 3 above.

13. H. B. Falls, *Exercise physiology* (New York: Academic Press, 1968).

14. Lucien Brouha, *Physiology in industry,* 2nd ed. (Oxford: Pergamon Press, 1967).

15. G. H. G. Dyson, *The mechanics of athletics* (London: University of London Press, 1962).

16. Vaughan Thomas, *Science and sport* (London: Faber & Faber, 1970).

17. A. W. Hill, *Trails and trials in physiology* (London and Beccles: William Clowes and Sons, 1965).

18. See references 3 and 4 above.

19. H. W. Knipping and A. Moncreiff, "The ventilation equivalent of oxygen," *Queensland Journal of Medicine,* vol. 25, 1932, pp. 17-30.

20. See reference 11 above.

21. W. C. Adams, "Influence of age, sex and body weight on the energy expenditure of bicycle riding," *Journal of Applied Physiology,* vol. 22, 1967, pp. 539-545.

22. Ibid.

23. J. Y. Harrison et al., "Maximizing human power output by suitable selection of motion cycle and load," *Human Factors,* vol. 12, no. 3, 1970, pp. 315-329.

24. See reference 7 above.

25. See reference 7 above.

26. P. O. Astrand and B. Saltin, "Maximal oxygen uptake and heart rate in various types of muscular activity," *Journal of Applied Physiology,* vol. 16, 1961, pp. 977-981.

27. R. C. Carpenter et al., "The relationship between ventilating capacity and simple pneumonosis in coal workers," *British Journal of Industrial Medicine,* vol. 13, 1956, pp. 166-176.

28. See reference 11 above.

29. Ian McDonald, "Statistical studies of recorded energy expenditures of man. Part II: Expenditures on walking related to age, weight, sex, height, speed and gradient," *Nutrition Abstracts and Reviews,* vol. 31, July 1961, pp. 739-762.

30. See references 1, 2, and 8 above.

31. G. A. Dean, "An analysis of the energy expenditure in level and grade walking," *Ergonomics,* vol. 8, no. 1, January 1965, pp. 31-47.

32. R. C. Garry and G. M. Wishart, "On the existence of a most efficient speed in bicycle pedalling and the problem of determining human muscular efficiency," *Journal of Physiology,* vol. 72, 1931, pp. 425-437.

33. See reference 29 above.

34. W. Brown, "Cycle gearing in theory and practice," *Cycling,* 5 July 1944 (London: Temple Press, 1944, pp. 12-13.

35. See reference 32 above.

36. F. R. Whitt, "Ankling," *Bicycling,* February 1971, pp. 16-17.

37. Ibid.

38. F. R. Whitt, "Pedalling rates and gear sizes," *Bicycling* March 1973, pp. 24-25.

39. M. J. A. Hoes et al., "Measurement of forces exerted on pedal and crank during work on a bicycle ergometer at different loads," *Internationale Zeitschrift für Angewandte Physiologie einschliesslich Arbeitsphysiologie,* vol. 26, 1956, pp. 33-42.

40 See reference 6 above.

41. See reference 21 above.

42. See reference 3 above.

43. See reference 31 above.

44. M. G. Bekker, *Theory of land locomotion* (Ann Arbor, Mich.: University of Michigan Press, 1962).

45. Velox, *Velocipedes, bicycles and tricycles: how to make and use them* (London: George Routledge & Sons, 1869).

46. See reference 31 above.

47. See reference 45 above.

48. "An experienced velocipedeist," *The Velocipede* (London: J. Bruton Crane Court, 1869), pp. 5-6.

49. See reference 13 above.

50. See reference 31 above.

51. Bill Bradley, "My Gross Glockner ride," *Cycling,* 25 July 1957, p. 90.

52. See reference 6 above.

53. Sylvia Dickenson, "The efficiency of bicycle pedaling as affected by speed and load," *Journal of Physiology,* vol. 67, 1929, pp. 242-245.

54. See reference 29 above.

55. See reference 31 above.

56. See reference 29 above.

57. S. W. Gouse, "Steam cars," *Science Journal,* vol. 6, no. 1, January 1970, pp. 50-56.

58. John Farey, *A treatise on the steam engine,* 1827. Reprinted by David Charles, London, 1971, p. 65.

59. See references 13-17 above.

60. See reference 16 above.

Additional recommended reading

Allen, J. G. "Aerobic capacity and physiological fitness of Australian men," *Ergonomics,* vol. 9, no. 6, 1966, pp. 485-496.

Astrand, I. "Aerobic work capacity of men and women with special reference to age," *Acta Physiologica Scandinavica,* vol. 49, Suppl. 169, 1960, pp. 1-92.

Chandler, N. R. and Chandler, C. H. "Tractive resistance to cycling," *Cycling,* 21 July 1910, p. B 2.

Hermans-Telvy and R. A. Binkhorst, "Lopen of fietsen? —kiesen op basis van het energieverbruik," Hart Bulletin, 6 June 1974, pp. 59-63.

Huxley, A. F. "Energetics of muscle," *Chemistry in Britain,* November 1970, pp. 477-479.

Judge, A. W. *The mechanism of the car* (London: Chapman and Hall, 1925), p. 180.

Moulton, Alex. "The Moulton bicycle," Friday-evening discourse, London, Royal Institution, 23 February 1973.

Sharp, A. *Bicycles and tricycles* (London: Longmans, Green & Company, 1896).

Shephard, R. J. "Initial fitness and personality as a determination in response to a training regime," *Ergonomics,* vol. 9, no. 1, 1966, pp. 1-16.

Wyndham, C. H. et al., "The relationship between energy expenditure and performance index in the task of shovelling sand," *Ergonomics,* vol. 9, no. 5, 1966, pp. 371-378.

3 How bicyclists keep cool

Bicycling can be hard work. It is very important that the body, like any engine, not become over-heated when producing power. We pointed out in Chapter 2 that the measurement of the power output of bicyclists on ergometers is open to criticism because the conditions for heat dissipation are critically different from those occurring on bicycles.

The performances of riding bicyclists in "time trials" are, however, very amenable to analysis. Such time trials are of far longer duration than the usual few hours assumed by Wilkie, for instance, as the maximum period over which any data are available for human power output.[1] Time trials (unpaced) are regularly held for 24-hour periods with, for instance, distances of 480 miles [772 km] being frequently covered (Table 2.4).

During bicycling the self-generated air blast is of such magnitude that it bears little resemblance to the draft produced by the small electric fans sometimes advised for cooling pedalers on ergometers. As a consequence it can be said that under most conditions of level running the riding bicyclist works under cooler conditions than does an ergometer pedaler. At high bicycling speeds most of the rider's power is expended in over-coming air resistance. At, say, 20 mile/h [8.94 m/sec] about 0.2 hp [149 watts] is dissipated in the air. The cooling is a direct function of this lost power. Even if the little fans often used for ergometer experiments ran at this power level, the cooling effect would be much less than that for the moving bicyclist, because little of this power is dissipated as air friction around the subject's body.

The effect of adequate cooling may be inferred from Wilkie's finding from experiments with ergometer pedalers[2,3] that if any capability of exceeding about half an hour's pedaling is required, the subject must keep his power output down to about 0.2 hp [149 watts]. However, peak performances in 24-hour time trials can be analyzed using data given in reference 1 on wind resistance and rolling resistance and this agrees with other published sources of similar information to show that some 0.3 hp [224 watts] is being expended over that period. It seems that the exposure of the pedaler to moving air is principally responsible for the improvement. It is also known that when a pedaler on an ergometer attempts a power output of about 0.5 hp [373 watts] he can expect to have to give up after some ten minutes and will be perspiring profusely. That is the same power output required to propel a racing bicyclist doing a "fast" 25-mile [40,233 m] distance trial involv- a duration of effort of nearly one hour. Again the striking difference produced by moving air upon a pedaler's performance is very apparent.

Let us examine the relevant literature for suitable correlations of established heat-transfer data in order to find quantitative reasons for the above observations.

Use of data on heat transfer

Because there is no published information concerning experiments on the heat transfer of actual riding bicyclists, it is necessary to make calculations with suitable approximations of a bicyclist's shape. The approximate forms used are a flat plate or 6-in.-diameter cylinder. In addition, data from experiments upon actual human forms can be looked at,[3,4,5] although the postures of the humans— lying flat or standing upright—were not those adopted by a riding bicyclist.

The results of many calculations using established correlations for both convective and evaporative heat transfer are given in Figure 3.1. Also shown is the heat evolution of a riding bicyclist at various power outputs and speeds on the level.

The figure indicates that the effect of shape upon the flux for a given temperature difference is not excessive in the case of convective heat trans-

Figure 3.1
Convective and evaporative
heat flows. Assumed conditions: surface temperatures
(constant), 35 °C; air
temperature (constant),
15 °C; air relative humidity,
80 percent.
Data for curves 1 and 2
from reference 7, p. 857.
Data for curve 3 from W.
H. McAdams, *Heat*

transmission (New York:
McGraw-Hill Book Com-
pany, 1942), p. 223.
Data for curves 4 and 8
from reference 4, p. 37.
Data for curve 5 from
reference 5, p. 257.
Data for curves 6 and 7 and
points 9 and 11 from ref-
erence 12, pp. 66, 69, 87,
88, 89.
Data for point 10 from C.

Strock, *Heating and ven-
tilating engineer's data-
book* (New York: Indus-
trial Press, 1948), pp. 5-12.
Data for curve giving heat
output of racing bicyclist
are from metabolic heat
data adjusted for mechanical power and some small
heat energy equivalents.
Bicyclists' body surface
assumed to be 1.8 m². See
Table 2.2.

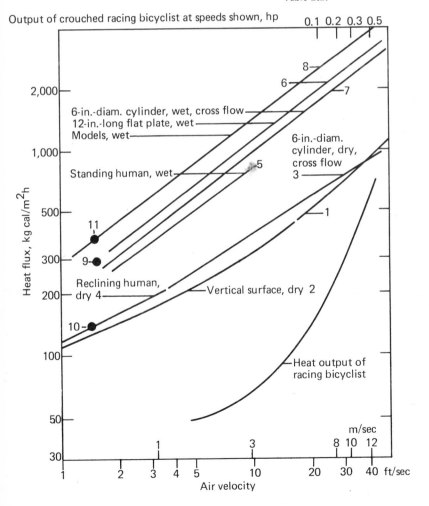

Output of crouched racing bicyclist at speeds shown, hp

fer. In the case of evaporative heat transfer the difference between results with models and an actual human body is 20 percent. It appears that a midway value can be obtained from data concerning cross flow over wetted 6-in.-diameter cylinders or flat plates. As pointed out by Colin and Houdas[6] for the same driving potential, expressed as water-vapor pressure or temperature difference, evaporative heat-transfer rates are about double those for convective heat transfer.

Deductions

Under normal free-convection conditions, data given in references 7 and 8 lead to the conclusion that we are being cooled by air moving at a velocity of about 1.5 ft/sec [0.457 m/sec]. This is supported in Figure 3.1, where line 6 for forced convection over a cylinder at 1.5 ft/sec [0.457 m/sec] and point 9 for free-convection conditions both predict about 280 kilocalories per square meter per hour as the heat flux for that air speed. This value for the air speed would be increased for bicyclists, because the pedaler's legs are also moving the air.

Information concerned with the design of heating and ventilating plants[9,10] shows that the maximum heat load produced by a hard physical worker has been long accepted as 2,000 Btu/h [586 watts]. This figure when applied to a man with a body surface of 1.8 square meters amounts also to 280 kilocalories per hour per square meter. It is recommended that such hard work should be carried out only at a room temperature of 55 °F [12.8 °C]. Most of this heat is lost by evaporation of sweat.

The evidence above leads to the conclusion that a man pedaling in such a manner that his body gives out a total of 2,000 Btu/h (586 watts) in average air conditions where free convection holds does not suffer from noticeable rise in body temperature, no matter how long the period for which he works. Because such a bodily heat loss for a pedaler on a stationary ergometer is associated with a mechanical power output of approximately

$$\frac{25}{100-25} \times \frac{2,000 \text{ Btu/h}}{33,000 \text{ (ft lbf/min)/h}} \times \frac{778 \text{ ft lbf/Btu}}{60 \text{ min/h}}$$

= 0.26 hp [194 watts] (25 percent thermal efficiency), it seems that a pedaler on an ergometer working for long periods produces only about 0.2 hp [149 watts] because he is unwilling to tolerate a noticeable rise in body temperature.

Earlier it was shown that many cyclists can exert 0.5 hp [373 watts]. According to Figure 2.5, that corresponds to a speed of about 27 mile/h or about 40 ft/sec [12.2 m/sec]. At that speed, the heat flow from the moving bicyclist is about 608 kilocalories per hour per square meter (Figure 3.1). If the cyclist exerts 0.5 hp [373 watts] pedaling on an ergometer, all the heat lost by convection and evaporation in moving air—all of the heat in excess of 2,000 Btu per hour—must be absorbed by the pedaler's body. Thus the ergometer pedaler with a body area of 1.8 m^2 absorbs

$$\frac{608 \text{ kcal/h/m}^2 - 280 \text{ kcal/h/m}^2}{60 \text{ min/h}} \times 1.8 \text{ m}^2$$

= 9.8 kcal/min

if the small heat losses through breathing are neglected.

If it is assumed that the pedaler's body weighs 70 kilograms and has a specific heat of 0.84 calories per gram per $^\circ$C, and that a rise in body temperature of 2 $^\circ$C is acceptable before physical collapse, the tolerable time limit for pedaling is:

$$\frac{70 \text{ kg} \times 0.84 \text{ cal/g·2}^\circ\text{C}}{9.8 \text{ kcal/min}} = 12 \text{ min}$$

From personal observations of highly trained racing bicyclists attempting to pedal ergometers at a power output of 0.5 hp [373 watts], a common range of endurance is 5 to 15 min. Hence the above estimations seem to be founded on sound theory. Such riders, incidentally, were all capable of racing in time trials of one-hour duration and more involving power outputs of nearly

0.5 hp [373 watts], thus vividly demonstrating the value of flowing air on the prolongation of the tolerable period of hard work.

Experimental findings supporting the foregoing are given in a paper by Williams et al.[11] concerning the effect of heat upon the performances of ergometer pedalers.

Conclusions

The heat-removal capacity of the air surrounding a working human is a key factor in deciding the duration of his effort. Static air conditions are apparently such that at low air speeds with free-convection conditions, the air is capable of removing 2,000 Btu per hour [586 watts] from the body surface of an average man. Hence if greater heat is given out from working at higher rates than about 0.2 hp [149 watts], body temperature rises. (A room temperature of 55 °F [12.8 °C] is assumed.)

The fast-moving air around a bicyclist traveling on the level can be estimated to have a heat-removal capacity much above that of the stationary air surrounding a man pedaling an ergometer. Quantitative estimations of an approximate nature can be made using established heat-transfer correlations based on air flow over wet 6-in [0.152 m]-diameter cylinders (cross-flow)[12] or from data given concerning air flow over a standing perspiring human.[13]

The heat-removal capacity of the air around a moving bicyclist at most speeds on the level is such that much more heat can be lost than that produced by the bicyclist's effort. Hence quite an amount of clothing can be worn compared with that tolerable to a static worker giving out the same mechanical power.

Some speculations

At least two ergometers used for testing the power capacities of racing bicyclists have incorporated air brakes in the form of fans. However, no one to date appears to have thought of directing the air from such air brakes on to the body of the pedaler and seeing what effect the fast-moving air had on the pedaler's performance. It is improbable that an air flow from such an arrangement could give anything very far from, say, half the flow rates surrounding an actual riding bicyclist giving out the same power. The results, however, would still be most interesting.

Pedaling on an ergometer out of doors should result in an advantage in the power of the pedaler. It is generally accepted that air movements around buildings at any rate are much faster than the 1½ feet per second [0.457 m/sec] quoted above for free-convection conditions around a heated body.

In view of the fact that at 0.2 hp [149 watts] output, for tolerable body temperatures, the body must get rid of its heat by an evaporative process, indoor exercise seems rather unhealthy compared with riding a bicycle in the open air. Maybe some of the exceedingly expensive home trainers sold for wealthy businessmen could be better designed by putting less into instrumentation and more into self-propelled cooling equipment.

**References
Chapter 3**

1. D. R. Wilkie, "Man as an aero-engine," *Journal of the Royal Aeronautical Society,* vol. 64, 1960, pp. 477-481.

2. Ibid.

3. T. Nonweiler, "Air resistance of racing cyclists," The College of Aeronautics, Cranfield, England, report no. 106, October 1956.

4. J. Colin and Y. Houdas, "Experimental determination of coefficient of heat exchanges by convection of the human body," *Journal of Applied Physiology,* vol. 22, no. 1, 1967, pp. 31-38.

5. D. Clifford, D. McKerslake, and J. L. Weddell, "The effect of wind speed on the maximum evaporative capacity in man," *Journal of Physiology,* vol. 147, 1959, pp. 253-259.

6. See reference 4 above.

7. J. R. Perry, *Chemical engineers handbook* (New York: McGraw-Hill Book Company, 1936), pp. 339, 958-965.

8. R. N. Cox and R. P. Clarke, "The natural convection flow around the human body," *Quest* (City of London University), 1969, pp. 9-13.

9. *Kempe's engineers year book,* vol. II (London: Morgan Brothers, 1962), pp. 761, 780.

10. O. Faber and J. R. Kell, *Heating and air conditioning of buildings* (Cheam, Surrey: Architectural Press, 1943).

11. C. G. Williams, et al., "Circulatory and metabolic reactions to work in heat," *Journal of Applied Physiology,* vol. 17, 1962, pp. 625-638.

12. T. K. Sherwood and R. L. Pigford, *Absorption and extraction,* (New York: McGraw-Hill Book Company, 1952), pp. 70, 87-89.

13. See reference 5 above.

Additional recommended reading

Martin, H. D. V. and Goldman, R. F. "Comparison of physical and physiological methods of evaluating the thermal stress associated with wearing protective clothing." *Ergonomics,* vol. 15, no. 2, 1972, pp. 337-342.

McAdams, W. H. *Heat transmission* (New York: McGraw-Hill Book Company, 1942).

Shephard, R. J. "Initial fitness and personality as determinants of the response in a training regime," *Ergonomics,* vol. 9, no. 1, 1966, pp. 1-16.

Strock, C. *Heating and ventilating's engineering databook* (New York: Industrial Press, 1948).

Whitt, F. R. "A note on the estimation of the energy expenditure of sporting cyclists," *Ergonomics,* vol. 14, no. 3, 1971, pp. 419-426.

The standard vertical riding position using pedals and cranks has evolved over the years into an accepted means for the satisfactory application of human power to moving a bicycle. However, there are some who believe that alternative mechanisms and motions might offer advantages. It seems worthwhile therefore to look at what evidence there is to see if this line of thinking has any value.

Early applications of muscle power to machinery

For thousands of years, prior to the advent in the 17th century of wind and water-mill power followed by steam and electric prime movers, man and animals had to be harnessed to provide mechanical power necessary for grinding corn, lifting water, and for other domestic or industrial work.

A common method of using man or animal power was to harness the walking man or animal to a revolving lever attached to a vertical axle (Figure 4.1). A more elaborate method was to let the man or animal walk either on an inclined disk (Figure 4.2) or inside a circular cage. An example of the latter type of "squirrel cage" can be seen in the Isle of Wight, UK, where at Carisbrooke Castle a donkey walks inside a wheel, which in turn moves a chain of buckets inside a well in order to raise water.

Tasks of a lighter nature than that of grinding corn or raising water in large quantities were performed with a hand-cranked handle motion (after the 8th century) followed by the use of the "bow" action in the Middle Ages (Figure 4.3). This latter action, where the foot alone was used, left the hands free to handle tools, for instance, on a lathe.

It appears that all the tasks associated with man-powered action in the earlier times were of a steady-motion character and were often ones involving heavy pushing rather than rapid limb

Figure 4.1
Horse-driven wheel.
Courtesy of Science
Museum, London.

movement. No rapid changes of speed were likely
to be required from an operator of such machinery.
Interesting accounts of the operation of these
types of machines can be found in references 1
and 2.

The coming of the man-propelled machine for road use

Although boats have been used over a long period
of time and have been moved by poles, paddles,
or oars, the man-propelled land vehicle dates back
only three hundred years. The propulsion of the
first man-moved land vehicles, which were three
or four wheelers, was achieved by limb actions
similar to those used for stationary machinery,
or occasionally by copying those used for punting
and rowing. This copying was reasonable, because
earlier methods were familiar and proved. The

Figure 4.2
Ramelli inclined footmill.
Reproduced with permis-
sion from *A history of
mechanical engineering* by
Aubrey F. Burstall
(London: Faber & Faber,
1963).

Figure 4.3
Bow-action lathe of the
Middle Ages. Courtesty of
Imperial Chemical Indus-
tries, Ltd.

vehicles could not be moved quickly, because
their resistance to rolling was great: roads were
rough and vehicles heavy. The proved propulsive
actions via levers, treadles, hand cranks, and the
like were appropriate to the circumstances. The
invention of the two-wheeled self-balancing machine by Macmillan in 1839 did perhaps introduce
an era of an easier-running machine, but the speeds
achieved were probably not outside the suitability
of the treadle system of propulsion.

Bicycle propulsion by means of pedals and cranks—a new system

According to Bury and Hillier[3] and Scott,[4] the
placing of cranks upon the front wheel of a bicycle is credited to Lallement of the Michaux concern who patented the idea when he arrived in
America (1866). It is recorded, however, that at
the 1862 Exhibition a tricycle was exhibited with
pedals and cranks on the front wheel.[5] This machine, made by Messrs. Mehew of Chelsea, pre-
dates the "Lallement patent" by several years
and it is probable that other inventors had thought
of the idea.* Even in the 1820s a cartoon had
appeared showing the Prince Regent in a position
(of some embarrassment) apparently turning,
with hand cranks, the front wheel of a "hobby-
horse."

---

*The hobby horse and boneshaker are two of the four
principal stages of evolution of the bicycle. Eighteenth-
century hobby horses were merely two wheels connected
by a board with a rudimentary seat (see Figure 9.1). They
were propelled by the "rider" pushing the ground with
his feet, and in the early models the front wheel could
not be steered. The boneshaker (1860s) had a steerable
front wheel with pedals (see Figure 9.3). In later boneshakers the front wheel was larger than the rear wheel
to provide better gearing, and the "ordinary" or "penny
farthing" evolved, with a large-diameter (around 60 in.)
[1.52 m] , pedaled front wheel and a small trailing
wheel (1870s). Many skull fractures and broken necks
resulted from spills from ordinaries, so that when the
development of the chain step-up transmission enabled
a bicycle to be made on which the rider could sit between
two wheels of equal moderate size, it was termed the
"safety" (1885). This is essentially the bicycle as we
know it today.

Never in the long history of man-moved machinery had the feet been used to push cranks. The bicycle was, even in its earliest days, relatively easy running compared with the heavy stationary machinery frequently turned by man's muscle power, and the introduction of a new propulsive system was welcome.

In spite of the rapid general acceptance that pedaling with cranks was a very efficient and practical method of bicycle propulsion, would-be inventors persisted in re-introducing centuries-old methods that imitated foot movements with very heavy stationary machinery. Scott illustrates many of the patents taken out by optimistic inventors from 1870 to 1890.[6]

More modern proposals for propulsion

There are several distinct methods of propulsion. Hand cranking is discussed first because it appears to be the one about which there is the most incontrovertible evidence as to its value compared with pedaling in the conventional manner.

Hand cranking: The power output of a human is dependent upon the duration of effort as well as upon the nature and conditions of the effort. Using evidence given in references 7 and 8 a series of curves has been drawn on Figure 1.6 showing how pedal power compares with that obtained from men working with winches or hand-operated ergometers. Although the circumstances of the tests are very different, the general conclusion is easy to reach: hand cranking is not a competitor for pedaling. Subsidiary evidence[9] also supports this view, because it shows that the maximum oxygen consumption of subjects cranking by hand is below that of the same men when pedaling. (Other experiments show that the oxygen-usage efficiency is very similar for action by both arms and legs so that efficiency is not a confusing factor.)

Pedaling is thus more suitable for human action to provide high power than is hand cranking. This conclusion is supported by the fact that although drivers of hand-propelled invalid carriages achieve remarkable performances, they in no way come close to, for instance, the performances of tricycle riders.

Hand cranking in addition to pedaling: Hand-crank mechanisms as additional means for applying muscle power to the standard pedaling arrangements date back a long time. In 1873 a patent was granted in the United States for an attachment to an "ordinary," and more recently several "safety" bicycles have been so fitted (Figure 4.4). Creditable performances have been claimed in spite of the acrobatic movements necessary for the rider to work hand cranks and steer simultaneously. Not all riders, however, could be expected to be so gifted.

Recent experimental work by Andrews with ergometers[10] has shown, rather surprisingly, that two muscle activities can be carried out simultaneously with a small gain in power for a given oxygen consumption. The matter has not been fully investigated as yet, or explained medically. It does appear, however, that for a particular set of circumstances, such as a need for an all-out high-speed effort, an additional hand-cranking mechanism has its virtues.

Pedaling in the horizontal position: Tests by Astrand and Saltin on ergometers have shown that pedaling in the near-horizontal position is only about 80 percent as effective, from the point of view of efficiency of muscle usage, as the normal upright position.[11] As far back as the 1890s the general advice to riders was "to get over the work." The riders of ergometers complain of knee strain when pedaling "sitting down." Information given by Haldane suggests that great strain is placed upon the muscular actions associated with breathing when a horizontal position is adopted.[12]

In spite of these conclusions, "sitting-down bicycles," particularly the 1930s "Velocar" type, have proved to be record breakers for *short* track

Figure 4.4
Hand-cranked bicycles.
(a) Sketch of bicycle on
which Fontaine broke the
London-to-York record in
1895. Courtesy of *Bicycling*
(b) Bricknell auxiliary
hand gear using rocking
handlebars. Harry "Goss"
Green broke many unpaced
records with this gear.

(a)

(b)

Figure 4.5
Reclining, hand-cranked
bicycle. Patent by I. F.
Wales, 2 March 1897.

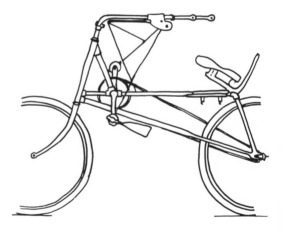

distances and to be acceptable to some touring-type riders (see Figure 5.3). It appears therefore that for some conditions the reduced wind resistance resulting from the lowered riding position outweighs the drawbacks of the apparently less advantageous use of the leg muscles. It is also possible that at high power levels, when the arms and back are required to counteract the pedal thrust in the upright position, the seat back in the recumbent bicycle acting directly on the hips reduces the otherwise useless stress on the upper torso. Semirecumbent bicycles have appeared at intervals. Examples of 1897, 1935, 1969, and 1972 are given in Figures 4.5, 4.6, 4.7 and 4.8.

Rowing action: A thorough and valuable investigation of human power output delivered by several types of rowing motion as well as by various crank actions was carried out by Harrison et al.[13] They pointed out that the usual rowing action is a motion in which, during each stroke, large masses of the body are given kinetic energy which has to be dissipated (and in boat propulsion the kinetic energy of the oars, and sliding seat if used, is also lost). Therefore, besides trying various subjects on an ergometer which could be arranged to give rowing motions in which the feet were held stationary and the body and seat moved, and in which the seat was stationary and the feet moved, Harrison investigated both "free" and "forced" motions.

Free motions were those in which the ends of the stroke were defined by the rower braking the kinetic energy—that is, the usual rowing action. In forced motions, the ends of the stroke were defined by mechanisms which turned the kinetic energy in one direction into energy in the reverse direction, in the same way as a pedal-crank system conserves energy.

As would be expected, the forced rowing motions gave substantially larger power than did the free motions. What was more interesting was that for all five subjects tested, the forced rowing motion with the seat fixed also gave more power

Figure 4.6
The Ravat horizontal
bicycle, 1935.

Figure 4.7
Captain Dan Henry's
recumbent bicycle. Courtesy
of Dan Henry.

Figure 4.8
Willkie recumbent bicycle.
Courtesy of Fred Willkie.

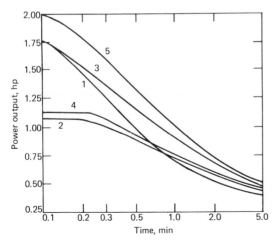

Figure 4.9
Human power by various
motions. Curve 1, cycling;
2 and 3, free and forced
rowing movements, respec-
tively, feet fixed; 4 and 5,
free and forced rowing, respectively, seat fixed.
From reference 13.

than did the normal pedal-cranking motion. Figure 4.9 gives the results for subject J. H. (presumably Harrison himself) showing a strikingly high peak-power level of 2 hp [1492 watts] for short-duration rowing.

Some circumstantial confirmation that screw-driven boats are more efficient than oar-propelled boats, and that pedaling gives more power than "free" rowing, is given by the performance of pedaled boats. When pedal-driven water cycles were in their heyday (1890s) the Thames was rowed by a triple-sculls boat during a 33-hour period; a speed some 18 percent greater was achieved by a "triplet" screw-propulsion water cycle. At about the same time other water cycles were proved to be quite speedy compared with normal boats. In particular a sextuplet water cycle ridden by six girls is alleged to have reached 15 mile/h [6.70 m/sec] on the Seine. This is a performance above that of racing eight-oar boats, rowed by good oarsmen.

Lever mechanisms: Re-introduction of the foot-pushed lever system is frequently proposed by those wishing to improve bicycle propulsion. Several conversions of bicycles to lever propulsion were carried out in the early 1900s. Levers were, of course, used on early bicycles in the mid-nineteenth century and on stationary machinery many centuries before (Figure 4.10).

In 1889 R. P. Scott came to the conclusion that for use on good roads at speed the normal pedal-and-crank system was excellent and mechanically less fragile than complicated lever systems.[14] He did, however, agree that some lever-and-clutch movements did give good hill-climbing attributes to bicycles. Hard and slow pushing is probably more efficient with levers than with rotary crank systems. In 1889 there were few variable-gearing mechanisms; when these appeared on cranked motions, pedaling at near-optimum rates could be used even during slow-speed ascents of steep hills. The pedal-and-crank system

Figure 4.10
The Macmillan steered,
self-balancing, lever-
driven bicycle of 1839

is also tolerably mechanically efficient compared
with a multijointed lever system and certainly
far better than any foot-pushed hydraulic-pump
with hydraulic-motor system as has been recently
proposed by some advocates of "improved"
propulsion methods.[15]

In modern conditions, particularly when
variable gears are available, it appears that the
lever system's alleged advantages for low-speed
heavy pushing can be bypassed and the normal
crank system, of known efficiency for high speeds,
used to advantage.

No ergometer experiments appear to have
been carried out on a lever-driven machine unless
they be by Harrison et al.,[16] but the results on the
muscle efficiencies associated with stepping and
walking up steep gradients are available.[17,18] Both
these leg actions are somewhat similar to the
thrust action of lever-propulsion systems. Experiments show that the muscle efficiency for pedaling
is in no way inferior to that associated with stepping and steep-grade walking (Figure 4.11). This
finding refutes the often proposed "theory" that
it is only by pushing the whole stroke vertically
that efficient usage of muscles is achieved and

Figure 4.11
Efficiency of various leg
actions. Net efficiency is
based on gross output less
resting metabolic output.

A: Gross efficiency,
pedaling.

B: Gross efficiency,
stepping.

C: Net efficiency, pedaling.

D: Net efficiency, walking,
40% (or 1 in 2.5) grade.

E: Net efficiency, walking,
30% (or 1 in 3.3) grade.

F: Net efficiency, walking,
5% (or 1 in 20) grade.

G: Net efficiency, walking,
20% (or 1 in 5) grade.
Data for curves A and B
from reference 17.

Data for curve C from
averages of reference 17
and "Report of the Bicycle
Production and Technical
Institute," 1968, Japan.
Data for curves D, E, and
F from reference 18.

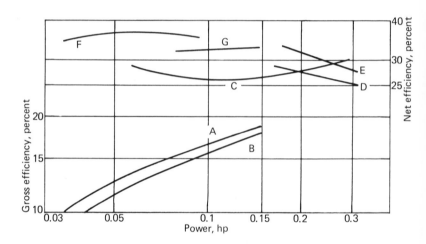

that the backwards-and-forwards foot movements
of pedaling over top and bottom dead centers
"wastes" energy.

Oval chainwheels. The zero torque at "top dead
center" is a factor in the design of engines as well
as bicycles. Under conditions of fixed gearing
and of heavy going, as were no doubt common
even on the level in the days of poor roads, the
problems of keeping the machine moving made
"getting over the top dead centers" a point of
vivid reality. One of the least complicated mechanisms invented for speeding up the foot at certain
points on the pedaling circle is the oval chainwheel.
This dates back to the 1890s.

The oval chainwheel was another mechanism investigated by Harrison et al.[19] All five subjects made steady-state power runs for as long as possible on the ergometer first with round and then with elliptical chainwheels. Four of the subjects showed no significant change of power output when they switched to the elliptic chainwheel. The fifth subject (J. H., again presumably the senior author) produced a somewhat higher power level, for short durations, when using the oval chainwheel rather than the round wheel (Figure 4.12). All subjects were accustomed to round chainwheels; although there was a training period for each mechanism investigated, it is possible that a longer period of training with the oval wheel might have shown improved performances with this device in all cases. In the 1930s the Thetic Company carried out some ergometer tests on their particular brand of oval chainwheel and claimed an appreciable benefit fot it, so that further testing would be justified.

Modern riding conditions and the use of variable gears result in riders now being less concerned with top-dead-center problems. Even "ankling,"* a technique much advocated in the early days of cycling, appears to be practiced less even by high-class racing men. Ankling is not an easy art for all riders to acquire, and the simple mechanism of the oval chainwheel is no mechanical embarrassment to the standard machine.

General conclusions

It appears that for all-round efficiency under the most commonly encountered circumstances the normal pedal-and-crank system with the rider in the vertical position is a well-proved method of human power generation. For particular circumstances when bursts of high speed are needed under favorable conditions of traffic or for record

---

*"Ankling" is the practice of bending the ankles in such a way as to maintain some thrust on the pedals during passage through the top and bottom dead centers of the pedal revolution.

attempts, a suitably trained rider could perform
better with various alternatives such as additional
hand cranking or by adopting a recumbent
position with the normal pedal-and-crank system,
or by a "forced" rowing motion.

Historical note

It appears from D'Acres writings of 1659[20] that
the squirrel-cage type of treadmill was considered
the most efficient type of manpowered engine
of that period. Foot-moved "treddles" apparently
could not "perform any great or worthy service"
and hand-operated winches or cranks needed the
assistance of "voluble voluntary wheels." This
latter term presumably described what is now
called a flywheel, a name reserved in the 17th
century for a type of fan brake (as is fitted to, for
instance, a clock still on show in Salisbury Cathe-
dral).

Figure 4.12
Power output with oval
(elliptical) chainwheel.
Data from reference 13.

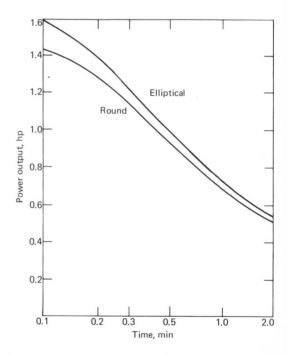

The favored treadmills were both large and power wasting through excessive frictional effects inherent in their design. (Far more modern horse treadmills of the 19th century were likewise considered very inefficient from the point of view of friction losses.) As a consequence, inventors of man-propelled carriages must have been placed in a quandary when they attempted to relate past experience with man-powered machinery to their designs. In early "manumotors" it was expected that the riders would have to push hard but slowly, an action common from the days of all man-powered machinery (see Figure 4.13, for example).

The appearance of the relatively easy running two-wheeled or even three-wheeled "boneshaker" vehicles brought about the need for a less forceful but faster and more variable type of man-to-machine connection. The crank action via the feet was for this purpose most appropriate in spite of the fact that it would not have been accepted in earlier times.

A useful feature of foot usage with pedals and cranks is that among the numerous muscle actions involved is that of the ankle movement which can assist the pedaling action either through the classic "ankling" method or, when toe-clips are fitted, through a "kicking forward" action at top dead centers. Walking also involves many leg motions with, however, more magnitude of swing of the heavy upper limb but with a helpful pendulum action to aid energy conservation.

A lack of rhythm is noticed when an experienced bicyclist tries a lever-driven machine, suggesting some equivalent of pendulum, or energy-conserving, action in the pedaling of rotating cranks. Some lever systems are also rather disconcerting in that there are no fixed limits to the length of stroke on the foot. The motion is equivalent to Harrison's "free" rowing action and the body must thereby both provide and dissipate kinetic energy in every stroke.

The early lever systems seem particularly appropriate to slow-speed, full-weight pedaling.

Figure 4.13
Medieval pump operated by
a treadwheel. Reproduced
with permission from *A
theatre of machines* by A.
G. Keller (London:
Chapman and Hall, 1964).

Figure 4.14
Harry Grant using curved
cranks while making a
paced record. Courtesy of
Harry Jelfs.

For modern bicycle riding but a fraction of a man's weight in thrust is necessary or even possible for other than brief periods of exertion.

A report on a "man-powered-land transport" competition in the British journal *Engineering* in 1968[21] shows that competitors were very interested in departing from the usual rotary system for application of man-power. However, no competitor satisfied the judges that he had basic experimental data upon which to justify his apparent enthusiasm for particular foot motions.

Curved cranks. This historical note would not be complete without a mention of a curious obsession of both early and late designers of machinery— the curved crank.

From the time of the introduction of the crank in about the 8th century, designers seem to have been equally divided as to whether it should be straight or curved. There is not "leverage" advantage in the shape of the curved crank; Keller[22] offers the explanation that users of the curved crank hoped for an extra motion to be derived from the curve. An additional phenomenon perhaps confused matters during the more recent centuries in that cast-iron wheels became common. Makers of these wheels found that if spokes were curved they were more flexible and less likely to break through differing rates of contraction during the cooling of the casting. Even in the 1930s, curved steel cranks had a following, and two first-rate track bicyclists were devotees. An earlier record breaker, Harry Grant, is shown using curved cranks in Figure 4.14. One could speculate that perhaps the use of curved cranks baffled a cycle-track opponent as to whether the user was able to jump with his pedal at dead centers or not.

References
Chapter 4

1. A. F. Burstall, *A history of mechanical engineering* (London: Faber and Faber, 1963).

2. R. D'Acres, *The art of water-drawing* (London: Henry Brome, 1659). Reprinted by W. Heffer and Sons, Cambridge, England, 1930.

3. Viscount Bury and G. Lacy Hillier, *Cycling,* third revised edition, The Badminston Library of Sports and Pastimes, London: Longmans, Green and Company, 1891.

4. R. P. Scott, *Cycling art, energy and locomotion* (Philadelphia: J. B. Lippincott Company, 1889), pp. 28-41.

5. See reference 3 above.

6. See reference 4 above.

7. E. A. Müller, "Physiological methods of increasing human physical work capacity," *Ergonomics,* vol. 8, no. 4, 1965, pp. 409-424.

8. J. C. Trautwine, *The civil engineer's reference book,* 21st edition (Ithaca, N. Y.: Trautwine and Company, 1937).

9. See reference 7 above.

10. R. B. Andrews, "The additive value of energy expenditure of simultaneously performed simple muscular tasks," *Ergonomics,* vol. 9, no. 6, 1966, pp. 507-509.

11. P. O. Astrand, and B. Saltin, "Maximal oxygen uptake and heart rate in various types of muscular activity," *Journal of Applied Physiology,* vol. 16, 1961, pp. 977-981.

12. J. S. Haldane, *Respiration* (London: Oxford University Press, 1922).

13. J. Y. Harrison et al., "Maximizing human power output by suitable selection of motion cycle and load," *Human Factors,* vol. 12, no. 3, 1970, pp. 315-329.

14. See reference 4 above.

15. David Gordon Wilson, "Man-powered land transport", *Engineering* (London), vol. 207, no. 5372, 11 April 1969.

16. See reference 13 above.

17. C. H. Wyndham et al., "Inter-and intra-individual differences in energy expenditure and mechanical efficiency," *Ergonomics,* vol. 9, no. 1, 1966, pp. 17-29.

18. Ian McDonald, "Statistical studies of recorded energy expenditure of man. Part II: expenditure on walking related to weight, sex, age, height, speed and gradient," *Nutrition Abstracts and Reviews,* vol. 31, July 1961, pp. 739-762.

19. See reference 13 above.

20. See reference 2 above.

21. See reference 15 above.

22. A. G. Keller, *A theatre of machines* (London: Chapman and Hall, 1964).

Additional recommended reading

Engineering Heritage, vol. 151 (Institution of mechanical engineers: Page Brothers, Norwich, 1963).

Bricknell, A. L. "The double-geared bicycle," Thames Iron Works Gazette, 30 June 1898, pp. 127-129.

Part II   Some bicycle physics

Wind resistance

The phenomenon of "wind resistance" is well known to everyone, and particularly to bicyclists, as an everyday experience. It is caused by two main types of forces: one normal to the surface of the resisted body—felt as the pressure of the wind—and the other tangential to the surface, which is the true "skin friction." For an unstreamlined body such as a bicycle and rider, the pressure effect is much the larger, and the unrecovered pressure energy appears in the form of eddying air motion at the rear of the body. Figure 5.1a shows this eddying effect at the rear of a cylinder particularly well. As can be seen in Figure 5.1b the streamlined shape produces less eddying than the cylinder.

Vehicles intended for high speeds in air are almost always constructed to minimize eddying. Streamlined shapes incorporate gradual tapering from a rounded leading edge. The exact geometry of shapes that maximize the possibility of the flow remaining attached (rather than eddying) and minimize the skin friction can be approximated by rather complex mathematics. It is usual in aeronautics either to refer to one of a family of published "low-drag" shapes or to test models in a wind tunnel.

Experimental investigations

The measurement of the wind resistance of motor vehicles is described by R. A. C. Fosberry.[1] Although good data in wind-tunnel experiments can be obtained for vehicles, better data can be given with mounted bicyclists because the interaction of the airflow around the bicyclist with the moving ground can be modeled more accurately than can the flow under and around an automobile.

One aim of aerodynamic experiments on an object is to measure its drag coefficient C_D, defined as:

Figure 5.1
Effects of bluff and
streamlined shapes.
(a) Eddying flow around
circular cylinder.
(b) Noneddying flow
around streamlined shape.
(c) Pressure recovery that
is possible in the absence
of eddies.

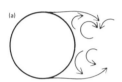

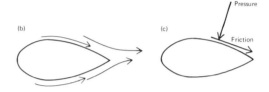

$$\text{drag coefficient} \equiv \frac{\text{drag force}}{\text{dynamic pressure of air} \times \text{frontal area}}$$

(nondimensional).

At low speeds (below, say, 50 mile/h [22.35 m/sec])
the dynamic pressure is given by:

$$\text{dynamic pressure} = \frac{\text{air density} \times (\text{relative velocity})^2}{2\,g_c}$$

where g_c is the constant in Newton's law $F = ma/g_c$
(see footnote on pages 23-24) and the relative velocity
is the velocity of the air moving past the object.
Thus the drag force is:

drag force = (drag coefficient $\times$ air density
$\times$ (relative velocity)^2
$\times$ frontal area)/2 g_c.

The propulsion power P necessary to overcome drag is:

P = drag force $\times$ relative vehicle velocity.

Since the drag force is proportional approximately to the square of the velocity, the power to overcome drag is approximately proportional to the cube of the velocity.

The vehicle velocity v is the same as the relative velocity used to calculate the drag force only in still air. When there is a head wind or a tail wind the relative velocity is different from the vehicle velocity.

If the drag is measured in pounds force and the velocity in feet per second the power is given in ft lbf/sec. This may be converted to horsepower by dividing by 550 (1 hp = 550 ft lbf/sec), or mile/h (1 hp = 375 mile lbf/h) may be used:

$$P\text{ (hp)} \equiv \frac{\text{drag (lbf)} \times \text{velocity (ft/sec)}}{550 \text{ (ft lbf/sec)/hp}}$$

$$\equiv \frac{\text{drag (lbf)} \times \text{velocity (mile/h)}}{375 \text{ (mile} \cdot \text{lbf/h)hp}} \; ; \tag{2}$$

in S. I. units, the relationship is

P (watts) $\equiv$ drag (newton) $\times$ velocity (m/sec) .

Because air density varies comparatively little at low altitudes, an approximate form of drag coefficient, k, has often been used, defined by:

$$k \equiv \frac{\text{drag (lbf)}}{\text{velocity}^2 \text{ (mile/h)}^2 \times \text{frontal area (sq ft)}} \; .$$

The drag force can be calculated as:

$$\begin{aligned}
\text{drag (lbf)} = \; & k \; [(\text{lbf/sq ft})/(\text{h}^2/\text{mile}^2)] \\
& \times \text{velocity}^2 \text{ (mile/h)}^2 \\
& \times \text{frontal area (sq ft)}.
\end{aligned} \tag{3}$$

The value of the constant k varies greatly according to the roughness of the sides of the body and relative length. Ordinary sedan automobiles have k values of 0.0015; racing-car values are about 0.0005. Railway locomotives have k values of about 0.002. A riding bicyclist has a k value of about 0.0023.

Both analysis and precise measurement show that the drag coefficient and the k value are not constants for any one vehicle or shape but vary slightly with velocity.

Drag coefficient values

Nonweiler[2] found that mounted bicyclists in racing clothing had drag coefficients C_D of about 0.9 where the average frontal area was taken to be about 3.6 sq ft (0.33 sq m) with the bicycle itself forming an appreciable proportion of the frontal area. Loose clothing increased the drag area by 30 percent. There is considerable independent evidence that 0.9 is a reasonable value for the circumstances. For instance, information referred to by Sharp[3] on the wind resistance experienced by bicyclists can be interpreted as being based upon a drag coefficient value of about 0.9. Wind-tunnel experiments on the upright human form, credited to A. V. Hill by Dean, give a value of about 0.9.[4] An account of aerodynamic work on the wind resistance of cylinders[5] is given by Rouse and can be interpreted as suggesting that an assembly of short cylinders, such as that representing the form of a bicyclist and machine, would have a drag coefficient of about 1.0.

It appears unreal to quote any value for these drag coefficients to greater accuracy than the first significant figure, because of the magnitude of the experimental errors involved.

Drag coefficients for other wheeled vehicles are given by Kempe.[6] The range is from 0.2 for sedan automobiles to 1.0 for square-ended motor trucks and to 1.8 for a motorcycle and rider. Racing cars have very low drag coefficients of 0.1 or less. Table 5.1 gives detailed information about these C_D values and Table 5.2 gives some estimates for "mopeds" based on published performance data. As would be expected, these are close to the values for bicycles and riders.

From the above deliberations emerges a numerical relationship between variables suitable for practical use with everyday units. It is assumed that the vehicles concerned are running at sea level when "standard" air density can be assumed. Then

Table 5.1 Values of C_D and k for formulas 1 and 3.

Drag coefficient C_D, nondimensional		$k, \dfrac{lbf}{ft^2} \cdot \dfrac{h^2}{mile^2}$
Sports car	0.2 - 0.3	0.00051 - 0.00077
Sedan car	0.4 - 0.5	0.00102 - 0.0013
Bus	0.6 - 0.8	0.00153 - 0.0020
Truck	0.8 - 1.0	0.0020 - 0.0026
Square plate	1.2	0.00307
Sphere	0.47	0.00120
Cylinder	0.7 - 1.3	0.0018 - 0.0033
Streamlined body	0.1	0.00026
Motor cyclist	1.8	0.0046
Racing cyclist	0.9	0.0023

Note: For average air conditions

$$k = C_D \times \frac{0.0765 \ lbm/ft^3}{2 \times 32.2 \ lbm \cdot ft/lbf \cdot sec^2} \times \left(\frac{88 \ ft/sec}{60 \ mile/h}\right)^2$$

$$= C_D \times 0.002555 \left(\frac{lbf}{ft^2} \cdot \frac{h^2}{mile^2}\right)$$

The density of the air is assumed to be 0.0765 lbm/cu ft and 88/60 is the conversion factor of mile/h to ft/sec.

Table 5.2. Air resistance of Mopeds.

Make	Engine power, hp	Weight, machine, lbm	Weight, rider, lbm	Max. speed, mile/h	Estimated data air resistance, hp	$k, \dfrac{lbf}{ft^2} \dfrac{h^2}{mile^2}$
Powell	1.05			26	0.64	0.0027
Mobylette	1.35	75	200	30	0.86	0.0024
Magneet	1.6	115	200	33	1.00	0.002

Note: Since force times velocity gives power, the power to overcome air resistance of an object with a frontal area of 5 sq ft is (see formula 3)

$$\text{air resistance (hp)} = \frac{k \ [(lbf/ft^2)/(h^2/mile^2)] \times velocity^3 \ (mile/h)^3 \times 5 \ sq \ ft}{375 \ (mile \cdot lbf/h)/hp}.$$

from the definition of the drag coefficient the following relation can be derived:

drag force (lbf) = $25.6 \times$ drag coefficient (C_D)
$\times$ frontal area (sq ft) $\times$ speed^2
(mile/h/100).^{2}

If bicyclists have a C_D value of 0.9 this takes the form:

drag force (lbf) = $0.0023 \times$ frontal area (sq ft)
$\times$ speed^2 (mile/h)^2.

Expressed in S. I. units the above is

drag force(newtons) = $0.043 \times$ frontal area (sq m)
$\times$ speed^2 (km/h)^2.

Streamlining: Complete streamlined casings have been used by racing bicyclists to raise their top speeds by about 6 mile/h [2.7 m/sec] over the usual maximum of 30 mile/h [13.4 m/sec] for particular events (Figure 5.2). From this information it can be concluded that the drag coefficient of these casings is about 0.25, which is credible because of the casings' resemblance to an enclosed automobile. Data given by Rouse show that above a certain "critical" velocity the air resistance of streamlined struts is considerably less than that of plain cylinders of the same frontal area.[7] The critical velocity depends on size and shape and is higher for frame tubing, for instance, than for the rider's body. It might be necessary to travel at an average of over 35 mile/h [15.6 m/sec] to make streamlining of the frame tubes worthwhile, whereas streamlining the body could pay off at much lower velocities.

Streamlining the tubing could reduce the wind resistance of the bicycle itself by a half at high speeds. Nonweiler suggests that the bicycle resistance could amount to about 1/3.6 of the total wind resistance.[8] If streamlining the tubes reduced the wind resistance by, say, ½, then the effect on total wind resistance (machine plus rider) would be 1/(3.6 × 2) or 1/7.2. A conservative view would be to take the reduction as 10 percent from the original wind resistance.

At racing speeds the power to propel rider and machine is almost all spent in overcoming air resistance, and this power is proportional to the speed cubed. If, therefore, the wind resistance is reduced by 1/10, the speed will have increased, for the same power, by approximately the cube root of (1 + 1/10). This ratio of speeds is 1.03 or a 3 percent increase in speed. Whether or not the rider thinks this worthwhile is a personal opinion. Records have, however, been broken with a speed increment smaller than 3 percent.

Tricycles

It has often been proposed that a tricycle with smaller-than-usual rear wheels could be faster than a conventional machine. If it is assumed that 16-in. [0.406 m] wheels can be used on a tricycle, the decrease in frontal area would be about 0.14 sq ft [0.013 sq m]. This is small compared with the average total area of man and machine, which is about 4.1 sq ft [0.381 sq m]. The area is actually reduced to about 0.96 of the original. The extra 4 percent power should therefore result in an increase of speed of 1.3 percent ($1.04^{1/3}$ is about 1.013). It could well be that some of this increase in speed due to lowered wind resistance would be lost because of the greater rolling resistance of smaller

Figure 5.2
Bicycle with streamlined enclosure. (Note that the design allows free circulation of air from beneath the rider, ensuring a cooling effect. See Chapter 3.)

wheels, although the stiffer wheels might counteract this in other ways. In any case, the possible
speed increase is very small and there appear to be
no really sound optimistic grounds for expecting
a small-wheeled tricycle to be faster than standard
large-wheeled-type tricycles.

Velocars

Another type of machine, the use of which can give
greater speed than the normal bicycle, is the velocar (Figure 5.3). The rider is seated feet forward
with the legs nearly horizontal. As a consequence,
the frontal area of rider and machine can be somewhat less than that of an "upright" bicycle. Moreover
the machine and rider are sometimes enclosed in a
more-or-less streamlined body. Information given
in a textbook *Fluid-dynamic drag* by S. F. Hoerner[9]
on the wind resistance of a man in various positions
suggests that a velocar with a seated rider should
experience about 20 percent less wind resistance
than a normal machine and rider. As a consequence,
for a given power output by the rider, the speed
should be several percent greater, assuming that
rolling and transmission losses do not greatly increase. This prediction has been borne out in
practice. In the 1930s, most short-distance track
records were broken by riders on velocars. It appeared, however, that for more prolonged periods
of effort the horizontal position of the rider tired
him more quickly and speeds achieved were no
longer any better than those on conventional bicycles. The riding position on a racing velocar

Figure 5.3
Racing "Velocar"—
recumbent bicycle.

forces the rider to press with his shoulders upon a rest, an action which wastes energy. Other approaches to streamlining bicycles are shown in Figure 5.4.

Figure 5.4
Some past attempts at streamlining bicycles. Courtesy of *Cycling*.

Effect of riding position on wind resistance

In this book whenever a typical example of a crouched racing bicyclist has been under discussion it has been assumed on the basis of evidence presented by Nonweiler[10] that the frontal area presented to the wind measures about 0.33 square meter. If a tourist-type bicyclist is under discussion (see Table 2.2) it has been assumed that the frontal area is about 0.5 square meter (these figures were used to calculate the curves A and B of Figure 2.6). The evidence for the 0.5 square meter lies in data presented by Sharp[11] and the senior author's (FRW) own experiments. The frontal area is obviously a function of the size of the rider, and bulkiness of his clothing, the bicycle and accessories, and in particular of riding position.

In this connection the wind resistance of skiers is relevant. Some interesting findings are given by Raines.[12] This experimental work has shown, for instance, that the position of the arms is of importance. In the "elbows-out" position appreciable extra resistance is experienced. The nearest approach of the skiing subject to that of a typical track bicyclist seems to be that of Figure 5.5. The resistance experienced at 50 mile/h [80 km/h] was a force of 20.5 lbf [91.3 newtons]. One could reasonably assume that the frontal area of the skier because of his accessories was near that of a crouched mounted bicyclist and machine. The drag force can be calculated as before:

drag force = $0.0023 \times 3.1 \times 50^2$ lbf

$$= 17.8 \text{ lbf } [79.2 \text{ newtons}].$$

The fairly close agreement of the estimate and the reported results is satisfying evidence that the data quoted in the previous discussion are realistic.

Aerodynamic forces on riding bicyclist caused by passing vehicles

All bicyclists when riding on roads frequented by fast and large motor vehicles have experienced side-wind forces from a passing vehicle.

No experimental work appears to have been reported concerning the magnitude of the lateral forces as far as actual bicyclists are concerned. Some most valuable work, however, has been

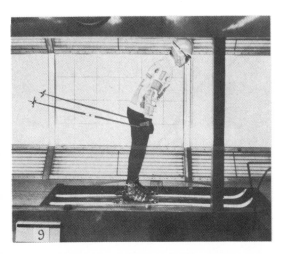

Figure 5.5
Aerodynamic drag of the
human body. Four positions
demonstrated by skier
Dave Jacobs were photo-
graphed in the NAL
tunnel at the same moment
that the drag was recorded.
The air speed was a steady
80 km/h. Standing erect
(run 9) Jacobs' drag was
22 kg. In a high but com-
pact crouch (run 15) the
drag was reduced by more
than half to 9.3 kg.
From reference 12.

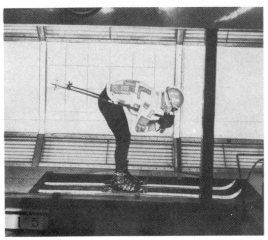

reported upon by Beauvais concerning the wind effects upon one-tenth-scale parked and jacked-up model automobiles.[13] Considerable concern exists in the United States about the safety of jacked-up vehicles situated at the side of expressways.

Interpreting Beauvais' data for bicyclists, we can estimate that bicyclists may experience lateral forces of several pounds when overtaken closely by large vehicles moving at 70 mile/h. The laws prohibiting bicycling along expressways are reasonable.

References Chapter 5

1. R. A. C. Fosberry, "Research on the aerodynamics of road vehicles," *New Scientist,* vol. 6, 20 August, 1959, pp. 223-227.

2. T. Nonweiler, "Air resistance of racing cyclists," The College of Aeronautics, Cranfield, England, report no. 106, 1956.

3. A. Sharp, *Bicycles and tricycles* (London: Longmans, Green and Company, 1896), p. 251.

4. G. A. Dean, "An analysis of the energy expenditure in level and gradient walking," *Ergonomics,* vol. 8, no. 1, January 1965, pp. 31-47.

5. H. Rouse, *Elementary mechanics of fluids* (London: Chapman and Hall, 1946), pp. 247.

6. *Kempe's engineers year book,* vol. II, (London: Morgan Brothers, 1962), p. 315.

7. See reference 5 above.

8. See reference 2 above.

9. S. F. Hoerner, *Fluid-dynamic drag,* (Midland Park, N. J., 1959).

10. See reference 2 above.

11. See reference 3 above.

12. A. E. Raine, " Aerodynamics of skiing," *Science Journal,* vol. 6, no. 3, March 1970, pp. 26-30.

13. F. N. Beauvais, "Transient aerodynamical effects on a parked vehicle caused by a passing bus," in Proceedings of the first symposium on road vehicles held in the City University of London, November 6 and 7, 1969.

Additional recommended reading

Shapiro, A. H. *Shape and flow* (New York: Doubleday, 1961).

The wheel and its rolling resistance

In the earliest times, man and animals moved only by means of leg motions applied via feet or hooves. Traveling by foot requires a several-fold variation in power for movement over hard, compared with very soft, ground, and walking can be said to be a reasonably adaptable means of locomotion. The resistance to the motion of a wheel, however, can vary several hundredfold from that on pavement to motion on soft soil. Hence, there was a real incentive to develop paved roads when wheels were adopted for horse-drawn vehicles (Figure 6.1). The ancient Roman empire was the first civilization to make use of this idea. It is recorded that the time taken to travel over Europe to Rome was less at that time than it was a thousand years later in the Middle Ages, when the Roman road system had vanished through lack of maintenance.

After the Middle Ages men overcame the stultifying effects of spiritual opposition to technological change, and inventions to improve man's everyday life rapidly appeared. Among these were iron-covered wooden railway lines, followed by iron wheels and cast-iron rails (1767). This gave rise to the Railway Age of Victorian times and was paralleled by a reappearance of a fair number of paved roads. Thomson, in 1845, followed by Dunlop in 1888, introduced pneumatic tires which decreased the rolling resistance of carriage wheels to nearer that experienced by the railway wheel and which also introduced a degree of comfortable riding on common roads. Thereafter, constant competition between the easy but "fixed" running of vehicles on steel tracks and the greater directional adaptability of road vehicles fitted with pneumatic tires has continued. It has been established beyond doubt that the minimum power to drive any practical man-made vehicle at a given constant speed is achieved by the use of steel wheels rolling on steel tracks. The power consumed in rolling the most flexible pneumatic-tired wheel is several times greater, and the average automobile wheel on the best surfaces generally available has ten or more times the resistance to motion of a railway-train wheel on its track.

Definition of the term "rolling resistance"

The power needed to propel wheeled vehicles depends not only upon the ease of rolling of the wheels themselves for a given set of conditions but also upon the physical properties of the surface. A great deal of information is available concerning the former in general and the latter for harder surfaces. Although wheel motion upon soft ground is of great interest to agricultural engineers and military-vehicle designers, this type of work is of less general interest. As a consequence, less information is available for the resistance offered by soft surfaces to the rolling of a wheel compared with that produced by hard roads.

The term "rolling resistance" as used in this book means the resistance to the steady motion of the wheel caused by power absorption in the contacting surfaces of wheel rim and road, rail, or soil upon which the wheel rolls. The power needed to accelerate or slow up a wheel because of its inertia is not included in the rolling resistance. The energy lost in acceleration is, for bicycle wheels, of small consequence compared with the power absorbed by tire and road: it does, unfortunately, often get referred to in the sense of "ease of rolling of wheels" and can be twisted into the statement that "little wheels roll more easily than large wheels." This latter is only partially true, even if "rolling" is taken to mean "accelerating and decelerating." (In steady motion on level ground, it does not matter how large the wheel is). As is discussed later, bicycle wheels are now of such a pattern that design changes can produce only small effects on acceleration properties, but a wheel of a given diameter has a rolling resistance, in the sense of surface-power absorption, of approximately only half that of a wheel of half this diameter. This

Figure 6.1
Replica of Egyptian chariot wheel of 1400 B.C. Note rawhide wrapping to make tire resilient. Reproduced with permission from the Science Museum, London.

type of rolling-resistance definition, as accepted in engineering literature, implies that the weight of rider and machine, both greatly exceeding that of the wheels, influences, via the tires, the motion of the bicycle; the rolling resistance (in, for example, lbf per ton) multiplied by the weight (in tons) gives the obstructing force.

The rolling resistance of railway-train wheels

The case of the rolling of a railway-train wheel has been thoroughly investigated.[1] It is more amenable to accurate measurement than are other wheel-rolling actions, such as that of pneumatic-tired wheels on roads. The hardnesses of the railway wheel and track can be specified closely and are less variable than other types of contacting surfaces.

The wheel rolling resistance is caused by the deformation of wheel and track producing a "dent" of a temporary nature, as shown in Figure 6.2. This deformation causes the point of instantaneous rolling of the wheel to be always ahead of the point geometrically directly under the center of rotation of the wheel about its bearing attached to the vehicle. The result is that a pair of forces which exert a retarding torque, known as a "couple," is set up. The numerical value of the torque is the downward force between wheel and surface, which in steady state is the weight of the wheel plus its share of the weight of the vehicle, multiplied by the distance $b/8$.

Koffman shows why the displacement of the instantaneous center of rotation can be calculated as the length b divided by 8.[2] Experiments have been carried out with railway-train wheels of typical diameters resting on rails and it has been found that the distance $b/8$ can be taken as 0.01 to 0.02 in. [0.254-0.508 m]. It is thus possible to calculate the rolling resistance according to the method given by Koffman. If the wheel radius is 20 in. [0.508 mm], the calculated rolling resistance is 1.1 to 2.2 lbf per long ton of vehicle weight [0.0048-0.0096 newtons/kq] on the wheel, in addition to bearing resistance.

Figure 6.2
Rolling-wheel resistance
diagram.

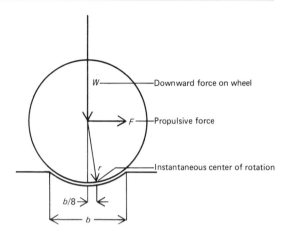

A check on the calculation given above can be carried out using information given in the *Engineering Encyclopedia* (p. F 532).[3] This source is relatively unique in that it gives a quantitative relationship for rolling friction of cylinders on plane surfaces:

$$\frac{\text{resistance}}{\text{to rolling (lbf)}} = \frac{\text{weight (lbf)} \times \text{coefficient of rolling friction, } f \text{ (ft)}}{\text{radius of cylinder (ft)}}$$

The experimentally determined values for f quoted include that for iron-to-iron contact with a range of 0.002 to 0.005. Substituting in the equation above the values of the weight as one long ton (2,240 lbf) and f as 0.002, we find for a 20 in. [0.508 m] wheel:

$$\text{resistance to rolling (lbf)} = \frac{2,240 \text{ lbf} \times 0.002 \text{ ft}}{20 \text{ in.}/12 \text{ in./ft}}$$

$$= 2.688 \text{ lbf}$$
$$[11.96 \text{ newtons}].$$

The rolling resistance of wheels on soft ground

The general effect of wheel form upon rolling resistance was investigated over a century ago by Grandvoinet.[4] He found that if the diameter of the wheel was increased 35 percent, the rolling resistance on soft ground decreased 20 percent. A similar increase in width decreased the rolling resistance by only 10 percent. For a very large wheel, it has been found that the tread width has a negligible effect on rolling resistance.[5] Other studies investigated the once-common wooden, steel-rimmed, agricultural wheel. The characteristics of modern pneumatic-tired military and agricultural vehicles are still being investigated. Not all concerned subscribe to the theory that these large-tired wheel vehicles can "float" on soil, as might be thought feasible.

In passing it is worth noting that a wheel driven on soft ground may require more effort than walking or running, which, whether associated with man or quadruped, are mechanisms of a different character. Races between bicyclists and runners over rough country show that the speeds of the two are much closer than for races on hard ground.

A great deal of experimental work has been carried out in more recent times on the power needed to move agricultural vehicles. Barger et al. describe some of this work[6] and examination of the original papers published describing the experiments in detail is very interesting. Barger and his co-workers have verified the general effects of wheel cross-sectional shape and diameter as postulated by the very early workers and have also carried out investigations on pneumatic tires. The main findings have been that wheel diameter, whether for a steel-rimmed wheel or for a pneumatic-tired wheel, is a most important factor. The larger the wheel, the more easily it runs when supporting a given weight, no matter whether the surface is soft or hard. For hard ground, the ease of running can be related to the diameter by a simple inverse-proportion formula; for soft ground the wheel-diameter effect is even greater.

When a loaded tire, pneumatic or steel, presses on a road surface, the shape of the area of deformation of the surfaces is much influenced by, among other things, the diameter of the wheel. If account is taken of the relative dimensions of the contact areas and reasoning along the lines employed for the railway-train wheel (see Figure 6.2) is used, it can be deduced that the forces opposing rolling are in fact inversely proportional to the wheel diameter. Readers interested in rolling-friction theory are advised to consult references 7 to 11 for further details about a subject which is not frequently referred to in textbooks on basic physics.

The rolling resistance of wheels fitted with pneumatic tires

The pneumatic-tired wheel rolling on the road exhibits exaggerated characteristics compared with the steel wheel on rails. For instance, the flattening of the tire over an "equivalent" distance b (see Figure 6.2) is obviously much greater for pneumatic tires, and therefore the theory predicts a much greater rolling resistance, as found in practice. What is very difficult to predict is the effect of flexing of the tire walls, which is so dependent upon inflation pressure and the design of the carcass, as compared with the constancy of steel's elasticity. An interesting peculiarity of pneumatic-tire rolling worth noting is that tires affect steering properties, because any side force applied to the wheel axle is resisted by the road at a point on the tire which is not directly beneath the axis[12,13] but slightly behind. This results in a measurable "twisting effect" not experienced by hard wheels on hard surfaces. This is called "self-aligning torque" and is a measure of the tendency of the steered wheel to follow the direction of motion. Tire-inflation pressure and carcass flexibility, obviously, also influence this twisting effect, as they do rolling resistance.

Early bicycles used solid rubber tires. The record times for the mile [1609.3 m] on the track for both the solid-rubber-tired "old ordinary" and the solid-rubber-tired "safety" are almost the same, both being close to 2½ minutes. It is known that the high bicycle offers greater wind resistance and needs more skill to ride than the smaller-wheeled bicycle. Hence, the findings above support the explanation that the bigger wheel runs more easily

than the smaller wheel[14] —the lower rolling
resistance compensates for the higher wind
resistance.

Although it might at first sight appear that
there are too many factors influencing pneumatic-
tire rolling for any simple correlation to be devised,
in practice this is not so. The predominant variables have been found to be tire-inflation pressure,
wheel diameter and road surface. Actual road
speed has an effect, but not until speeds well
above those common for bicycles are involved is it
appreciable.[15] For modern bicycles running on hard
roads, the range of each of the three predominant
variables is only about twofold, giving a total
possible effect of some eightfold on the rolling
resistance.

**Quantitative measure-
ment of the rolling
resistance of pneumatic
tires**

As stated above, the rolling resistance of pneumatic
tires is a combination of several resistances, not
all of which can be predicted theoretically. Experiments, nowadays generally using towed wheels,[16]
have therefore to be carried out in order to measure the force in lbf/long ton [or newtons/kg] of
vehicle necessary to move it under various circumstances. The data given by these experiments are
discussed further below.

Formulas for calculating the rolling resistance
of automobile tires of about 5 in. (127 mm) cross
section are given by Bekker[17] and Kempe.[18] Some
information concerning bicycle tires of 1¼ to 2 in.
[31.75 mm to 50.8 mm] cross section has also
been reported by Patterson[19] and Sharp.[20] All information shows that the most important factor
influencing ease of rolling is that of the inflation
pressure of the tire, presuming that road surface,
wheel size, and cross section are similar. It seems
probable that the rolling resistance of a bicycle
tire on a wheel 26 in. [660.4 mm] diameter, on
smooth roads, ranges from 22 lbf/long ton supported [0.0963 newton/kg] to about 12 lbf/ton
[0.0525 newton/kg] if the inflation pressure is
varied from 17 lbf/sq in. [1.172×10^5 newton/

m^2] —at which pressure the rim is liable to "bump" on the road and give warning of gross misuse to the careless rider—to the 75 lbf/sq in. [5.17×10^5 newton/m^2] recommended by tire makers. Less smooth or hard surfaces, such as rough macadam or gravel, may cause an increase of 50 to 100 percent. For a given roughness of the surface and a given load, the larger the wheel, the easier it rolls, a fact also established over centuries by experience in the field of horse-drawn vehicles.

Examination of quantitative information on tire rolling resistance

The earliest accessible information on the bicycle tire seems to be that given by Sharp[21] (see Table 6.1). Three values for the coefficient of friction of tires on road and track are quoted from a publication by C. Bourlet.[22] No tire pressures are specified, although it was well known by then that this factor has a major influence on the ease of rolling of tires. Patterson carried out more recent (1955) experiments,[23] which are summarized in convenient form in Table 6.2. Two formulas for calculating the rolling resistance of automobile tires, given by Bekker[24] and Kempe,[25] are quoted later in this chapter.

Table 6.1. The rolling resistance of early tires.

| Road surface | Rolling resistance, lbf/ton | | Speed, mile/h |
	Solid tire	Pneumatic tire	
Racing track		8.96	
road[a]		11.2 - 22.4	
Road, smooth			
macadam[b]	50 - 60	30 - 35	
Flag pavement[c]	60	33	5
Flint[c]	60	31 - 37	4 - 10

Sources:
[a]Cycle tires; data from reference 20, p. 251.
[b]Car tires; data from A. W. Judge, *The mechanism of the car,* Vol. III (London: Chapman and Hall Ltd., 1925), p. 150.
[c]Heavy cycle tires; data from reference 20, p. 256. It is probable that the high figures quoted for these entries are due to the investigator, H. M. Ravenshaw, including the air resistance, in addition to rolling resistance, in his results.

Available data on the effect of the tire pressure and wheel diameter on rolling resistance are combined in Figure 6.3. Because no tire pressures are quoted for the information credited to Bourlet,[26] it has been necessary to assume that appropriate limits are 55 to 80 lbf/sq in. [3.792×10^5 to 5.516×10^5 newton/m^2]. The two formulas quoted later predict similar values for C_R and but little effect from vehicle speed in the low range of speeds, applicable to cycling, of up to about 12 mile/h [5.36 m/sec]. (If the curves had, however, been calculated for 30 mile/h [13.41 m/sec] C_R values would have been increased by only a few percent.) These formulas and others are discussed at length by Ogorkiewicz,[27] who also stresses the applicability of curve-A data (in Figure 6.3) and other predictions from the formula, even in present-day car design, although the basic experimental work was carried out in Germany almost forty years ago when tires were of different construction from those of the present. It is most probable that the wheel diameter used was similar to that of modern bicycles, 26 to 27 in. [660 to 683 mm].

Table 6.2. Experimentally determined tire rolling resistances.

Tire, in.	Load, lbf	Speed, mile/h	Tire inflation, lbf/in.^{2}	Rolling resistance, hp	Rolling resistance R, lbf/ton
2	120	20	10	0.1	36
2	120	20	18	0.07	25
2	120	20	30	0.05	18
2	150	20	18	0.1	28.5
2	180	20	18	0.12	28.5
1¾	120	15	18	0.05	23
1¼	120	15	45	0.02	9.7

Source: Reference 19, pp. 428-429.

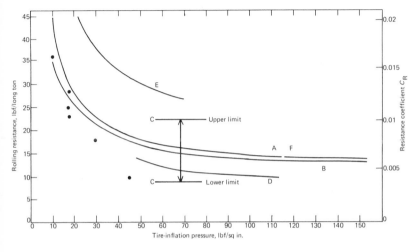

Figure 6.3
Effect of tire-inflation pressure on rolling resistance.

Curve	Wheel	Surface
A	auto	smooth, hard
B	auto	smooth, hard
C(limits)	bicycle, 28 in. X 1 1/2 in. (ave.)	road and track
D	bicycle, 27 in. X 1 1/4 in.	smooth, hard
E	bicycle, 16 in. X 1 3/8 in.	medium rough, hard
F	bicycle, 27 in. X 1 1/4 in.	medium rough, hard
• (points)	bicycle, 26 in. (assumed) X 1 1/4 in.	steel rollers

Data for curve A from reference 18 (bias-ply tires).
Data for curve B from reference 4 (bias-ply tires).
Data for curves D, E, and F from experimental data by Whitt (for low speeds).
Data for points • from reference 19 (see Table 6.2).

The senior author (FRW), with the help of several other bicyclists riding several different bicycles and tricycles on typical roads frequented by bicyclists, has carried out experimental work on rolling resistances of tires.[28] All tires were of 1 1/4 or 1 3/8 in. [31.75 or 34.92 mm] cross section and of light construction. The total weight of rider and machine was always near 180 lbm (81.65 kg). The experiments showed that for concrete or rolled-gravel surfaces the rolling resistances were very close to those predicted by curve A of Figure 6.3. This means that light bicycle tires on rough surfaces have lower resistance coefficients than the larger-cross-section automobile tires—in other words, bicycle tires do not require as good a road surface for a given performance. The results quoted by Patterson[29] also show that bicycle tires roll more easily than car tires. Information, in general, suggests that the performance predicted by curve D can be attained on first-class hard roads by 1 1/4 in. [31.75 mm] cross-section light bicycle tires.

Experiments with small-wheeled bicycles showed that, as predicted by Barger et al.[30] working with pneumatic-tired tractors, the rolling resistance is increased in near proportion as the wheel diameter is decreased for a given constant inflation pressure. The small-wheel "low-pressure" big-cross-section tire is the slowest both because of the small diameter of the wheel and the designed low inflation pressure of the tire (35 lbf/sq in. [2.41×10^5 newton/m^2]).

For comparative purposes, Figures 6.4 from Ogorkiewicz[31] and 6.5 from a report of the Motor Industry Research Association[32] are included to show how little speed affects the rolling resistance of car tires, although tire-pressure effects are appreciable in the speed range 30-50 mile/h [13.41-22.35 m/sec].

Table 6.3 has been included to show how great is the rolling resistance of steel-tired wheels on roads compared with that of pneumatic tires inflated to high pressure. No doubt this fact became immediately apparent to riders of the early "boneshakers." These machines, in their latter days of usage, were often manufactured with rubber tiring attached to their wheels, in a manner adopted for many years afterwards by makers of horse-drawn carriages. Hollow, square-section rubber tiring was also used as well as solid tiring, even as early as 1870.

The use of information on tire rolling resistance

Information given by curve D of Figure 6.3 and that on wind resistance given in Chapter 5 has been used to compile Tables 6.4 and 6.5. These tabulations show how tire pressures affect the rate of movement of a bicyclist under various conditions. In particular, the table shows a predicted 5 to 10 percent slowing effect, for a given power, of the tricycle's extra wheel and axle compared with a bicycle. This prediction is substantiated by the times achieved in records for the two types of machines. The effect of the use of good solid-rubber tires is also revealed in Table 6.5 in which the rolling resistance is about the same as that of a pneumatic tire at about 12 lbf/sq in. [0.827 $\times 10^5$ newton/m^2] pressure, that is, about 30 lbf/ton [0.131 newton/kg] of vehicle weight (see Table 6.1). This should be of general interest to

Figure 6.4
Effect of inflation pressure on automobile-tire rolling resistance. From reference 15.

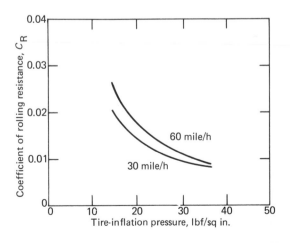

Figure 6.5
Effect of speed on
automobile-tire rolling
resistance. Each point is
the mean of 6 measure-
ments, and the standard
deviation is indicated. Tire
size, 5.50 x 16; load,
720 lbf. rolling resistance
coefficient =

rolling resistance
_____
load

(a) Variation of rolling
resistance with tire pressure
and speed. Road surface,
Tarmac.
(b) Variation of rolling
resistance with road surface.
Pressure, 30 lbf/sq in. From
MIRA report (reference 16).

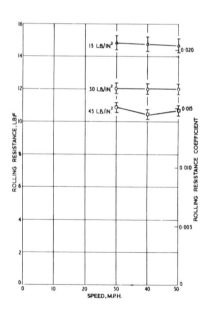

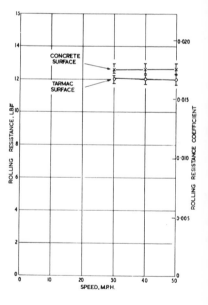

Table 6.3. Rolling resistance of four-wheeled wagon (steel tires) and 1½ ton stagecoach.

Road surface	Rolling resistance R, lbf/ton	Speed	Vehicle
Cubical block Pavement	32 - 50	slow	wagon
Macadam	62 - 75	slow	wagon
Planks	30 - 50	slow	wagon
Gravel	140	slow	wagon
"A fine road"	76 - 91	4 - 10 mile/h	stage coach
Common earth road	200 - 300	slow	wagon

Source: Reference 7, p. 683.
Note: Reference 3 cites work with 24 to 60-in. diameter steel-rimmed wheels and gives
R values of a similar range.

Table 6.4. Total rolling resistance calculated. (Load 170 lbf, curve D, Figure 6.3).

Speed, mile/h	Tire press. (1¼ in.), lbf/sq in.	Rolling resistance, hp	Rolling resistance, lbf/ton	Total power lost (includes air resistance), hp	Wheel diam., in.	Percentage speed reduction for given power due to small wheel
30	75	0.070	11.5	0.69	27	
30	17	0.140	23	0.767	27	
29.3	75	0.113	19	0.69	16	2.3
12.5	75	0.029	11.5	0.074	27	
12.5	17	0.058	23	0.103	27	
11.4	75	0.044	19	0.074	16	8.8
5	75	0.0116	11.5	0.0140	27	
5	17	0.0233	23	0.0265	27	
3.6	75	0.0138	19	0.014	16	28
9.8	35	0.0337	17	0.053	16	

riders of old bicycles and tricycles who are made aware, forcibly, of the slowing effect of solid-rubber tires.

The power needed to overcome rolling resistance is given by

power (hp) = rolling resistance (lbf/ton) × weight (tons) × speed (mile/h)/375 (mile lbf/h)/hp

or, in S. I. units,

power (watts) = rolling resistance (newton/kg) × weight (kg) × speed (m/sec).

Unlike the power needed to overcome wind resistance, which is proportional to the speed cubed, the power lost in rolling is directly proportional to the speed, at least at low speeds.

If a bicyclist had only rolling friction to overcome, it can be estimated from tire formulas that he should attain speeds of over 100 mile/h [44.7 m/sec] on good surfaces. World records for bicyclists riding behind fast cars indicate that as much

Table 6.5. Effect of tire pressure on propulsive power needed.

Vehicle	Load, lbf	Speed, mile/h	Tire press. (1¼ in.), lbf/sq in.	Rolling resistance, hp	Rolling resistance, lbf/ton	Total power lost (includes air resistance), hp	Percent increase to total power (75 lbf/sq in. pressure "standard" 27 in. wheel)
Cycle	170	25	75	0.059	11.5	0.407	
Cycle	170	25	17	0.118	23	0.466	14
Cycle	170	12.5	75	0.0295	11.5	0.074	
Cycle	170	12.5	17	0.059	23	0.103	39
Tricycle	180	23.5	75	0.082	17.2	0.407	
Tricycle	180	23.5	17	0.164	34.8	0.489	20[a]
Tricycle	180	13.4	75	0.476	17.2	0.11	
Tricycle	180	13.4	17	0.0952	34.8	0.157	43
Cycle	170	25	N.A.	0.154	30	0.508	25[b]
Cycle	170	12.5	N.A.	0.078	30	0.122	65[b]

[a]Note tricycle is 6 percent slower than bicycle.
[b]Solid tires, 5/8 in. diameter.

as 120 mile/h [53.6 m/sec] can be attained for
short distances, thus verifying the estimation. (It
is arguable that air friction is not merely brought
to zero, but may actually help to propel a rider
pedaling behind a moving shield.) It is probable
that a runner, shielded from the wind in a like
manner, would improve his performance of a maximum of about 20 mile/h [8.94 m/sec] only
slightly, because air-friction effects for a runner
are relatively low compared with the other resistances at this speed.

On referring, in addition, to Figures 1.2 and
2.5, some other interesting conclusions can be
drawn. For instance, at maximum bicycle speeds,
if the bicycle had no friction or mass and only its
air drag resisted motion, the top speed would increase by only a few percent. At low speeds, the
situation is rather different: at about 10 mile/h
[4.5 m/sec] such a machine would require about
half the power needed from the rider under normal
conditions. If the same power were to be exerted
on a weightless, frictionless machine, the speed
would be increased by about 30 percent, to 13
mile/h [5.8 m/sec].

Advantages and disadvan-
tages of small-wheeled
bicycles

In recent times there have appeared on the market
new bicycles incorporating wheels of 14 to 20 in.
diameter [355 to 508 mm], compared with the
common diameter of 26 or 27 in. [660 or 686
mm]. This design feature appears to be accepted
as essential if the bicycle is to be easily stowed in
the trunk of a car and if one machine is to be safely
ridden by people of different heights. In addition,
luggage can be carried more easily over a smaller
wheel simply because there is more space available.
And some designers have incorporated springing
into small-wheeled bicycles. It appears that these
requirements are considered to be important for
those of the general public who may be deterred
for various reasons, both sociological and practical,
from using a conventional machine.

A question often raised about small-wheeled
machines is the effect of the smaller wheels on the

Figure 6.6
Effect of tire pressure and wheel diameter on propulsive power required for bicycles. Note the diminished rate of decrease of power required at pressures above 75 lbf/sq in., the manufacturer's recommended pressure.

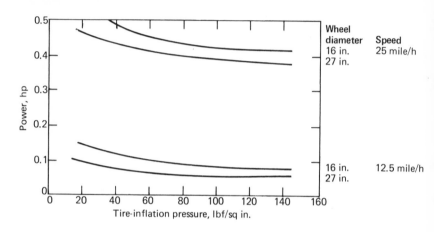

Figure 6.7
Slowing effect of 16-in.-diameter wheels compared with use of 27-in.-diameter wheels at same power level. Note: the 27-in. wheels are assumed to be running on a smooth road surface with a rolling resistance of 11.5 lbf/long ton weight, and the 150 lbm rider is crouching and has a frontal surface area of 3.65 sq ft. The drag coefficient is 0.9. The percentage drop in speed for a "slower" machine, that is, with a rolling resistance of 18 lbf/long ton and with a frontal area of 5.5 sq ft, is not very different. Point ● is a single estimation for such conditions. In both cases the tire pressure is 75 lbf/sq in.

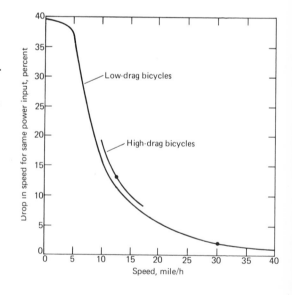

propulsive power needed from the rider. The extent to which this power, for a given rate of progress under specified conditions, exceeds that needed by a conventional machine depends, of course, on other details of the particular bicycle design, as well as on wheel size. Of great importance is the tire inflation pressure at which the machine can be ridden with comfort. "Soft" tires add resistance for all sizes of wheel, as was discussed earlier in this chapter, whether the low pressure is one deliberately intended by the designer or is due to the rider lacking the strength of arm (or memory) to reach a desirable inflation pressure (about 55-60 lbf/sq in. [$3.79 \times 10^5 - 4.14 \times 10^5$ newton/m^2] for 26-27 in. diameter [660-686 mm] 1 3/8 in. [34.92 mm] tires. The effect of inflation pressure on rolling power for two wheel sizes is shown in Figure 6.6.

We have estimated the rolling and air resistances for a popular size of wheel of 16 in. [406.4 mm] diameter and compared the power requirements at different speeds with those for 27 in. [686 mm] diameter wheels, and the results are shown in Tables 6.4 and 6.5 and Figure 6.7. These calculations have been drawn up to show the calculated quantitative effect of the use of different tire pressures and wheel diameters on the power needed for riding on very good roads. It is obvious that the smaller wheels are "slower" over the whole range of speeds, and to an appreciable extent at the lower speeds. (If rougher roads had been assumed for the calculation, the "slowness" would have been more apparent—unless the wheels were assumed to be incorporated in a sprung, damped suspension, when they can be superior.) At speeds of 25 to 30 mile/h [11.18 to 13.41 m/sec] and higher, the effect of the smaller wheels is relatively small, according to the calculations, because wind effects are predominant. This accounts for the experience in practice that racing times for the 27 in. [686 mm] wheeled machines are closely approached by the smaller-wheeled machines.

Whether or not the appreciable slowing of the smaller wheels at utility and touring speeds of 10 to 12 mile/h [4.47 to 5.36 m/sec] is acceptable depends, of course, upon the temperament of the rider.

The rolling resistance R may be calculated by the methods of references 33 and 34:

$$R = C_R m,$$

where m is the weight of machine plus rider and C_R, the coefficient of friction, is given by

$$C_R = 0.005 + \frac{1}{p}\left[0.15 \frac{\text{lbf}}{\text{sq in.}} + 0.35 \frac{\text{lbf h}^2}{\text{sq in. mile}^2}\right.$$
$$\left. \times \left(\frac{\text{speed (mile/h)}}{100}\right)^2\right],$$

where p is inflation pressure, lbf/sq in., for 27-in.-diameter wheels.

The coefficient of friction C_R multiplied by 2,240 gives the rolling resistance R in lbf per long ton of vehicle.

Smooth treads on automobile tires reduce rolling resistance by as much as 20 percent according to information given by Ogorkiewicz.[35] He and Bekker[36] give an alternative formula for calculating C_R.

$$C_R = 0.0051 + \left[\frac{0.0809 \text{ lbf/sq in.} + 0.00012\, m/\text{sq in}}{p\,(\text{lbf/sq in.})}\right.$$
$$+ \left[\frac{0.105 \text{ lbf h}^2/\text{sq in. mile}^2 + 0.0000154\, m}{p\,(\text{lbf/sq in.})}\right]$$
$$\left. \times \left[\frac{V\,(\text{mile/h})}{100}\right]^2\right.,$$

where m is weight on wheel, lbf.

The effect of wheel mass on riding effort required for acceleration

The wheels of a vehicle move both forward with the machine and rider and at the same time rotate around the hubs. The resistance of the wheels to a change in speed is therefore greater, per unit mass, than that offered by the rest of the vehicle. Hence, greater effort is required to accelerate "a pound of weight (mass) in the wheel of a bicycle than a pound in the frame." This fact has been quoted endlessly in cycling literature, both in and out of context.

The wheels of a bicycle are now of a form such that the major portion of the mass is concentrated in the rim, tire, and tube combination. The dimensions of the latter are small compared with the diameter of the wheel and their center of mass is close to the outside of the wheel, which is traveling at road speed. On this account, it is possible to say with some truth that "the effect of a given mass in the wheels is almost twice that of the same mass in the frame" as far as acceleration power requirements are concerned, because the wheel has to be given both the translational kinetic energy of the whole machine, and its own rotational kinetic energy relative to the bicycle.

With modern bicycle construction the wheels form only about 5 percent of the total mass of machine and rider. Also, the effect of any practical variation in reducing this 5 percent is small, whether by reducing the wheels by size or by material content. At the best, it is estimated that the wheel mass can be reduced to 3½ percent of the total. The reduction effect is therefore a 5 minus 3½ or 1½ percent. Even if this can be multiplied by two because the mass revolves, the resultant 3 percent effect on acceleration is very small and would not be easy to detect.

More accurate estimations based upon calculations or measurements of the actual moments of inertia of 16-in. wheels compared with 27-in. wheels show that the difference in acceleration power is rather less than 1.7 percent.

Although the lighter wheels accelerate slightly more quickly for a given power, and have a lower air drag, they also have a larger rolling resistance on smooth roads, because of the larger losses at the point of contact (see Figure 6.2). The decision on whether or not to use small wheels obviously must depend on the duty anticipated for the bicycle, as well as on cost and fashion.

Figure 6.8
Dynamics of wheel losses
on rough surfaces.

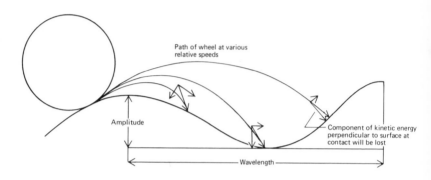

Path of wheel at various
relative speeds

Amplitude

Component of kinetic energy
perpendicular to surface at
contact will be lost

Wavelength

**Rough roads and
springing**

Rough roads affect bicyclists in several ways. The
vibration may be uncomfortable and may require
the bicycle to be heavier than if it were designed
for smooth roads. And there will be an energy
loss.

The energy loss depends on the "scale" of the
roughness, the speed, and on the design of the
bicycle. If the scale is very large so that the bicyclist has to ride up long hills and then to descend
the other side, overall energy losses are small (and
principally due to the increased air-resistance losses
at the high downhill speeds). There are in this
case virtually no momentum losses.

Now imagine a very small scale of roughness,
with a supposedly rigid machine traveling over the
surface. Each little roughness could give the machine an upward component of velocity sufficient
for the wheel(s) to leave the surface (see Figure
6.8). The kinetic energy of this upward motion
has to be taken from the forward motion, just as

if the rider were going up a hill. But when the wheel and machine descend, under the influence of gravity as before, the wheel contacts the surface at an angle, the magnitude of which depends upon the speed and the scale of the roughness. All the kinetic energy perpendicular to the surface at the point of contact can be considered to be lost. Herein lies part of the reason for rough-road losses.

Pneumatic tires greatly lower the losses for small-scale roughness because only the kinetic energy of part of the tread is affected, and the spring force of the internal pressure ensures that in general the tire does not come out of contact with the surface. The principal losses are due to the flexing of the tires and tubes ("hysteresis" losses).

At a larger scale of roughness, perhaps with a typical wavelength of 6 to 60 in. (0.152 to 1.52 m) and a height amplitude of 1 to 6 in. (25 to 152 mm), bicycle tires are too small to insulate the machine and rider from the vertical velocities induced, and the situation more nearly approaches the analogy to the rigid-machine case discussed above. For this scale of roughness, typical of pot holes and ruts, some form of sprung wheel or sprung frame can greatly reduce the kinetic-energy or momentum losses by reducing the unsprung mass and ensuring that the wheel more nearly maintains contact with the surface.

Another way of expressing this conclusion is that, if energy losses are to be small, the "natural" frequency of the unsprung mass should be high compared with the forced vibrational frequency imposed by the surface. The natural frequency f_N of a mass m connected to a spring having a spring constant λ (λ gives the units of force applied per unit deflection) is

$$f_N = \frac{1}{2\pi} \sqrt{\frac{\lambda g_c}{m}} \text{ vibrations per unit time.}$$

The forced frequency from the road surface is

$$f_f = \frac{v}{S},$$

where v is the velocity of the bicycle and S is the wavelength of the roughness. Therefore the ratio

$$\frac{f_N}{f_f} = \frac{S}{2\pi v} \sqrt{\frac{\lambda g_c}{m}}$$

should be kept high by reducing the unsprung mass m for the worst combination of S and v thought likely to be encountered. (The designer has little choice for the spring constant λ because he must assume a mass of rider and machine of up to perhaps 275 lbm [124.7 kg] (having a weight at sea level of 1223 newton) with a maximum deflection, if a light rider is to be able to reach the ground with his foot, of perhaps 3 in. (76 mm).

Road and track bicycles

Throughout this book the motion of the bicycle under consideration has been assumed to be taking place upon relatively smooth surfaces. In such circumstances it seems reasonable to assume that energy losses due to vibration are small. Roads are certainly becoming smoother. As a consequence the task for bicycle designers has been made easier than it was in the earlier days when even in the industrialized societies most of the roads were too rutted for easy riding. In the United Kingdom, where much sporting activity in the cycling world is carried out in the form of time trials, the modern road-racing bicycle is approaching the track bicycle in detail design, as with, for example, small-cross-section lightweight tires. Present-day utility machines are little different in specifications from road racers of the 1920s, another sure indication that much bicycle riding can be done upon good roads.

Opinions of early bicyclists

In contrast to the foregoing, the pre-1890s bicycle designer was forced to take serious account of the road surface in the road/machine combination. An early writer (see Scott[37]) was of the opinion that if the front wheel of a rear-driving safety (fitted with solid rubber tires) was forced to surmount a 4-inch-high obstacle, a loss of one-half of the forward momentum was experienced. This is an extreme case but is indicative of the large energy losses likely when riding on very rough roads. It was also known that solid rubber tires were less easy running as the speed increased even upon relatively smooth roads; the vibration loss is almost directly proportional to speed, even at low speeds. According to Sharp,[38] C. Bourlet, the French engineer, thought that one-sixth of the rider's effort was lost to vibratory effects when riding a solid-rubber-tired bicycle.

Early antivibration devices

As can be expected with the above state of affairs, prevailing inventors busied themselves with so-called antivibratory devices of all imaginable types.

Satisfactory designs for the application of antivibration mechanisms to bicycle frames were found most difficult to make. Several designers seemed to have a clear grasp of the essential problems to be solved: the rider must not have to cope with differing distances between saddle and pedals and forward momentum must be preserved. The general outcome was, however, far from being optimum, and Scott comments, "the difficulty experienced by inventors on the line of anti-vibrators appears to be, that while acquiring the desired elasticity in the proper direction, an elasticity in other directions has followed, making the machine feel unsteady and capricious, especially in the steering. This undoubtedly valid difficulty in the way is worthy of careful consideration before accepting an anti-vibrator: in fact the very desired end can be easily missed in an imperfect device, as it might, while holding momentum in one direction lose it in another."[39]

In spite of difficulties, inventors persevered and there was some sale for machines fitted with a large antivibrator (as distinct from sprung forks or saddles) in the form of a sprung frame. Three examples are shown in the figures. The type of frame most praised was the "Whippet" (Figure

Figure 6.9
The Whippet spring-frame
bicycle. From reference
20, p. 296.

6.9). All machines suffered from the effects of
wear of the joints, to varying degrees, and what
might have been an acceptable machine when new
was not so when the joints became loose. The
steering of the "Whippet" pattern is seriously affected by wear, as can be surmised by even a
casual inspection of the design. Practical riding
experience is both enlightening and awe-inspiring
when gained upon a sprung frame which is loose
in its essential joints.

The final deliverance out of the sufferings,
both mental and physical, of those concerned was
through the invention of the pneumatic tire in
1888. This invention placed the antivibratory device just where inventors had always wanted it,
at the road surface, thus doing away with a chain
of actuating connections to the root of energy
absorption. At first the pneumatic tire was almost
impractical because of its proneness to cutting by
road litter. Rapid development proceeded, and by
1892 most new bicycles were sold with pneumatic
tires, although the cost was very high compared
with solid rubber or hollow rubber tires (called
cushion tiring). An interesting warning is given in
an early text on bicycles.[40] This says that the
pneumatics of the contemporary design were prone
to roll on cornering and thus could cause fear to

Figure 6.10
The Humber spring-frame
bicycle. From reference
20, p. 297.

the less intrepid riders. Maybe this fact and the
fragility of the tire delayed its universal acceptance
among nonracing riders by a year or two. It must
be emphasized that for road use the early pneumatic tires appeared to be run at an inflation pressure of 20-30 lbf/sq in. [1.38×10^5 to 2.07×10^5
newton/m^2], which is far too low for cornering
with ease of mind. It was probable that it was
thought advisable to avoid strains due to high inflation pressures, which could have split the covers,
although on the other hand it could be assumed
that puncturing was made more easy through the
use of such low pressures.

The designer of the "Whippet" frame is thought
to have been convinced that there was no future
in large-scale adaptation of springs to bicycles after
the date of the introduction of the pneumatic
tire. These sentiments were not shared by other
innovators, however, and we see that even the
large Humber concern thought that there was a
demand for a sprung frame, though pneumatic
tires were fitted to the bicycle (Figure 6.10). Over
the following decades this example was followed
by others incorporating pneumatic and other unusual springing, some of which may have been
inspired by the design of light motorcycles which
appeared in the 20th century. No doubt for very

rough roads such sprung machines could have been useful, but the average road conditions for bicycle riding were getting better and thus decreasing the need for major springing devices in bicycles.

In the less-developed parts of the world, where bicycles are ridden in quantity, the roads are still rough. The most common bicycle is one fitted with large-diameter tires of about 28 in. by about 1½ in. [about 700 mm by 39 mm] cross section. This ensures a tolerable riding comfort without the resort to a sprung frame.

The Moulton design

The appearance of a successful modern design of sprung bicycle would seem to contradict the above arguments. However, the logical reasoning of the designer, Alex Moulton, was as follows. For bicycles to be truly useful to the "utility" bicyclists there has to be better provision for the carrying of luggage than can be fitted to standard machines. If wheels were much smaller, room for luggage carriers over the wheels would be created. Small wheels would lead to unacceptable vibration and energy losses, especially with "dead" loads (the luggage) over them, so that sprung wheels are required. Small wheels also make the bicycle a little shorter, so that it can fit into the trunk of a European standard automobile. The rear-wheel spring uses rubber in compression and shear, and the front wheel has a coil spring with rubber for damping (Figure 6.11). The resulting bicycle is very effective over both smooth roads and over those too rough for regular bicycles to tackle at any but very low speeds.

Dan Henry's sprung lightweight

A very successful though noncommercial design of sprung bicycle is shown in Figure 6.12. This has been developed by Captain Dan Henry of Flushing, N. Y., as a modification of a lightweight sports machine. Each wheel is mounted in a swinging fork on stiff bearings, thus maintaining lateral rigidity while giving long up-and-down travel. The springs are quickly adjustable for rider weight. The

Figure 6.11
Moulton bicycle, with
front-wheel springing.
Courtesy of Raleigh
Industries, Inc.

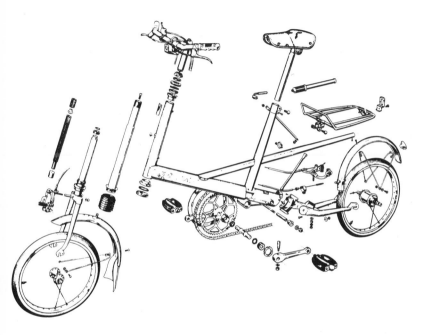

Figure 6.12
Captain Dan Henry's
spring-frame bicycle.
(a) Detail of the front
suspension.
(b) Normal position.
Courtesy of Dan Henry.

wheelbase is lengthened from that of the standard
machine because of the rearward placing of the
rear wheel, but the steering geometry is unaltered
(with the front wheel in its mean position) through
the setting back of the original front forks to
compensate for the forward set of the swinging
forks.

Dan Henry has ridden over 100,000 miles on
this machine, which weighs 28 lbm [12.7 kg]. He
notes two features of his experience which are con-
trary to those quoted for other sprung bicycles. He
finds that he is faster in hill climbing than on an unsprung machine. And his tires last longer; he is
able to use lightweight sew-up tires on roads
where clincher (wired-on) tires would be necessary
with unsprung bicycles.

References
Chapter 6

1. J. L. Koffman, "Tractive resistance of rolling stock," *Railway Gazette* (London) 6 November 1964, pp. 889-902.

2. Ibid.

3. *Engineering Encyclopedia* (New York: Industrial Press, 1954).

4. See M. G. Bekker, *Theory of land locomotion* (Ann Arbor Mich.: University of Michigan Press, 1962), pp. 209, 214.

5. E. Barger et al., *Tractors and their power units* (New York: John Wiley and Sons, 1952).

6. Ibid.

7. J. C. Trautwine, *The civil engineers reference book,* 21st edition (Ithaca, N. Y.: Trautwine and Company, 1937).

8. See reference 3 above.

9. J. Hannah and M. J. Hillier, *Applied mechanics* (London: Sir Isaac Pitman and Sons, 1962), p. 36.

10. Osborne Reynolds, "Rolling Friction," *Philosophical transactions,* vol. 166, 1876, pp. 155-156.

11. I. Evans, "The rolling resistance of a wheel with a solid rubber tire," *British Journal of Applied Physics,* vol. 5, 1954, pp. 187-188.

12. V. Steeds, *Mechanics of road vehicles* (London: Illiffe and Sons, 1960).

13. P. Irving, *Motorcycle engineering* (London: Temple Press 1964), p. 10.

14. See reference 5 above.

15. R. M. Ogorkiewicz, "Rolling resistance," *Automobile Engineer,* vol. 49, May 1959, pp. 177-179.

16. G. M. Carr and M. J. Ross, "The MIRA single-wheel rolling-resistance trailers," Motor Industries Research Association, Nuneaton, Warwickshire, England, 1966.

17. See reference 4 above, p. 209.

18. *Kempe's engineers year book,* vol. II (London: Morgan Brothers, 1962), p. 315.

19. P. D. Patterson, "Pressure problems with cycle tires," *Cycling,* 28 April 1955, pp. 428-429.

20. A. Sharp, *Bicycles and tricycles* (London: Longmans, Green and Company, 1896), p. 251.

21. Ibid.

22. C. Bourlet, *La bicyclette, sa construction et sa forme* (Paris: Gauthier-Villars, 1889), pp. 85-97.

23. See reference 19 above.

24. See reference 4 above.

25. See reference 18 above, p. 313.

26. Quoted in reference 20 above.

27. See reference 15 above.

28. F. R. Whitt, "Power for electric cars", *Engineering* (London), vol. 204, no. 5296, 2 October 1967, p. 613.

29. See reference 19 above.

30. See reference 5 above.

31. See reference 15 above.

32. See reference 16 above.

33. See reference 18 above, p. 315.

34. S. F. Hoerner, *Fluid-dynamic drag* (Midland Park, N. J., 1959).

35. See reference 15 above.

36. See reference 4 above, p. 208.

37. R. P. Scott, *Cycling art, energy and locomotion* (Philadelphia: J. B. Lippincott Company, 1889).

38. See reference 20 above, p. 252.

39. See reference 37 above.

40. Viscount Bury and G. Lacy Hillier, *Cycling,* third revised edition, in *The Badminton Library of Sports and Pastimes,* London: Longmans, Green, and Company, 1891.

Additional recommended reading

Kamm, W. Gesamtfahrwiderstandsgleichung für die fahrwiderstande von personenkraftfahrzeugen, 1938, DKF ZB 24.

Moulton, Alex. "The Moulton bicycle," Friday-evening discourse, London, Royal Institution, 23 February 1973.

7

Resistances to motion due to mechanical friction

Chain-transmission power losses

The retarding effects of wind, road, and gradient have been discussed in previous chapters. Another, but far less important, resistance to the progress of a bicycle rider is that due to friction-power absorption by the chain transmission and the bearings of the modern machine. No estimates for these pedal-power requirements have been included in Figure 1.2 or Figure 2.5.

The loss of power in an automobile transmission can be as high as 15 percent according to reference 1. This loss occurs principally in the gear reduction and the idling pinions in the transmission and differential, both sets of gears being oil-immersed and operating at relatively high speed. The efficiency of a good clean chain can be as high as 98.5 percent according to references 2 and 3. The loss of only 1.5 percent is very small in comparison to the power consumption of the wind and road resistances opposing bicycle motion. For example, at a speed of 12.5 mile/h [5.59 m/sec] (see Table 6.5) when a power of 0.074 hp [55 watts] is needed to overcome both wind and road resistance, only 0.001 hp [0.75 watt] is absorbed by the transmission. The tire rolling resistance (0.0295 hp [22 watts]) cannot be estimated to this degree of accuracy (0.001 hp in 0.0295 hp, or 3 percent), let alone the power absorbed by the wind. It appears reasonable, therefore, to refrain from including machinery losses in graphs of power usage for bicycle riding as exemplified by Figures 1.2 and 2.5.

In the early days of bicycle construction, there was a preponderance of machines with front-wheel drive, which was to be expected because of the simple, lightweight, and 100 percent efficient transmission of power from the pedals. The disadvantages, however, are serious when speeds are higher than the few miles per hour of the earliest days of cycling. The wheel must be made as large as possible, reaching the 60-in. (1.52 m) size of the high "old ordinary," to give high "gears." This, along with the limited steering arc of the wheel and the need for applying a torque to the handlebar to resist the pedaling torque, made the machine difficult for the less acrobatic to master. The addition of gear trains or the use of levers complicated the inherently simple type of drive and made it less attractive on this account. Lever drive may have some advantages at relatively low speeds, but for speeds now commonly possible with bicycles, the rotary motion of pedals on cranks seems physiologically sounder than a straight up-and-down motion. The reason might be the lack of control at the ends of the stroke of many lever mechanisms, which has been shown by Harrison et al.[4] to result in a power production lower than that achieved with rotary pedals. However, Harrison found that a fully controlled linear motion could deliver more power than any other type he investigated (as discussed in Chapter 2).

Some details of the evolution of modern chain design are given in references 5 and 6-10. Chain-driven bicycles were first used on very rough roads. This environment, along with the Victorian passion for manufacture in cast iron, appeared to influence chain and chainwheel design. "Open link" chains with thick and wide teeth on the cogs (partly because of the low strength of cast iron) were common practice. It was said that the road grit dropped more easily through the big spaces between the links. The small number of teeth led to rough running because of the variation in the speed of the chain (as much as 6 percent) in passing over a constant-speed cog.

In later times gearcases (oil-bath chain and cog enclosures) became common, even for racing machines, until the roads improved. Smaller-pitch chains then came into use, with improved running characteristics: there was only about 1 percent variation in speed with constant-speed drive. The precise shape of teeth has been subject to much experiment, and a modern opinion on the optimum design, credited to Renold, is given by Charnock[11] and Kay.[12] This design uses an angle of 60 degrees between the flat faces of two teeth, with circular arcs to the root of the teeth and also to the tips. The exact nature of these curves is, however, even now the subject of much discussion from the point of view of world standardization, and technical committees have not yet agreed on the general policy.[13]

Illustrations of gearwheel and chainwheel teeth shown on advertising posters, even for engineering exhibitions, are frequently open to criticism for the "artistic license" used in the production of ugly, inoperable shapes for teeth. The mechanically fastidious reader will be pleased to know that all artists do not escape effective criticism. It is recorded that the celebrated French poster artist Toulouse Lautrec once lost a commission for drawing a bicycle advertisement because his illustration of a chain set and chain was outrageously incorrect in the eyes of the manufacturer sponsoring the advertisement.

Power absorbed by bearing friction

For a long time, since at least 1896, the retarding effect of the friction of the standard ball bearings of a bicycle has been considered very small. Sharp[14] (p. 251), quotes Professor Rankine as stating that the friction forces amount to one thousandth of the weight of the rider. For a 150 lbf [68 kg] rider this means 0.15 lbf [0.667 newton] resistance or 1.8 lbf per long ton [7.9×10^{-3} newton/kg] for a rider on a 30 lbf [13.6 kg] bicycle. The tire rolling resistance (let alone wind resistance) is not known to be better than about 0.1 percent of the vehicle weight; hence it appears reasonable to disregard the bearing resistance. It is, however, interesting to compare this 1.8-lbf-per-ton [7.9×10^{-3} newton/kg] estimate with later relevant information, such as that for railway rolling stock as given in references 15 and 16. The wheel plus bearing rolling resistance is given there as a few lbf per ton. Of this the bearing friction alone is probably under 2 lbf per ton [8.8×10^{-3} newton/kg] of vehicle weight. The power loss in the complete transmission of an ergometer machine was given by Wilhelm von Döbèln[17] as being as low as 5 percent. According to references 1 and 2, chain power losses probably average 2½ percent. The bearing losses can thus be taken as 5 — 2½ percent. At power inputs to a bicycle of 0.12 hp [89 watts] and 0.37 hp [276 watts] (representing speeds on the level of 12 and 20 mile/h [5.36 and 8.94 m/sec] for a touring-type machine with an upright-seated rider) the total opposing forces can be calculated as 3.75 and 7 lbf [16.68 and 31.14 newton], respectively. The frictional opposing force of 0.15 lbf [0.67 newton] given by Rankine is thus expressible as 0.15/3.75 $\times$ 100 or about 4 percent at 12 mile/h, and 0.15/7 $\times$ 100 or about 2 percent at 20 mile/h, and the average of these two cases is 3 percent.

Information given in Table 7.1 shows that the coefficient of rolling friction attributable to 1-in. diameter [25.4 mm] balls in a bearing is about 0.0015, and Levinson shows that the rolling friction varies inversely as the ball diameter.[18] In addition, we can assume that the typical average angular contact of the bearing is 45 degrees, thereby increasing the effective load on the races by $\sqrt{2}$, or 1.41. Hence the average bicycle ball bearing (1/4 to 1/8 in. [6.35 to 3.175 mm] balls) should have a coefficient of 0.0015 $\times \sqrt{2}$ /(3/16) = 0.011. Experimental data are given in Figure 7.1.

The friction force for wheel bearings is very small because of the leverage effect of the relatively large wheel. The power losses in the bracket, pedals, and rear-wheel bearings considered as part of the transmission can be estimated from the above coefficient of friction as rather less than 3 times 0.011 $\times$ 100 or about 3 percent.

It appears from the above evidence that the

Table 7.1. Bearing friction.

Type of bearing	Coefficient of friction
Ball bearing (annular)	0.00175[a] 0.0005 - 0.001[b] 0.001 - 0.0015[c] 0.0015[d]
Roller bearing (small needle)	0.005[c]
Plain metal (gun metal- good lubrication)	0.002 - 0.015[e]
Machine tool plain metal slow running fast running	 0.1[f] 0.02[f]
Nylon 66 Dry-on nylon on metal Lubricated	0.04[g] 0.2[g] 0.07[g] 0.14[g]
P.T.F.E. (Glacier type)	0.1 - 0.14[h]
Ball bearing (bicycle type)	0.01[i] 0.01 average[j] (radial or thrust loads)

Sources:
[a]1-in. balls; data from R. P. Scott, *Cycling art, energy and locomotion* (Philadelphia: J. B. Lippincott Company, 1889) p. 175.
[b]1-in. balls; data from reference 18, supplement II.
[c]Reference 1, vol. I, p. 1242.
[d]Reference 3, p. 48.
[e]Reference 3, p. 49.
[f]*Mechanical world year book* (Manchester, England: Emmott & Co., 1938), p. 442.
[g]*British Plastics,* February 1966, p. 80.
[h]Manufacturer's leaflet: The Glacier Metal Company, Ltd., Alperton, Wembley, Middlesex, England.
[i]A. Sharp, *CTC Gazette,* October 1898, p. 493. Efficiency data of Professor Carpenter.
[j]Author's (FRW) own experimental work with 1/4 to 1/8 in. balls in angular contacts of 30° and 60°, 3/4 in.-diameter running circle.

Note: The bicycle-type bearings are assumed to be in very good condition and carefully adjusted. Otherwise the friction can be several-fold greater. Likewise poor manufacture can give such variations.

Figure 7.1
Test results for bicycle ball
bearings.
Line A: Rolling friction
coefficient, 0.0015.
Line B: Rolling friction
coefficient, 0.001.
Both lines are for a 1-in.-
diameter ball with an
angular thrust of 45°.

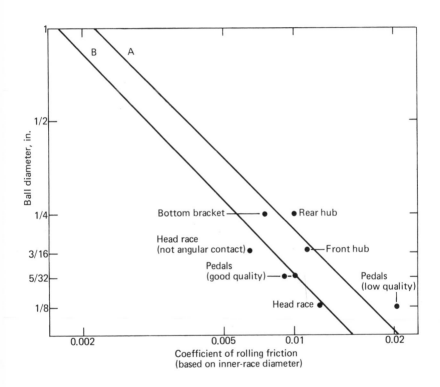

power losses in bicycle bearings of machines in good condition is only a few percent of the total power used.

Advantages of ball over plain bearings

The first ball bearings were far from being the highly reliable product we expect today, but they were very soon adopted by bicycle makers. Only a few years elapsed before the plain bearings of the "boneshaker" period were abandoned in favor of the more complicated ball bearings of the then contemporary design. Patents for ball bearings intended for bicycle use were more numerous than for other purposes in the early period of bearing history.

The common cup-and-cone bearing, which is inexpensive and can tolerate some degree of mis-alignment—a very desirable characteristic for in-accurately made and somewhat flexible bicycle construction—appeared as early as in the 1880s (Figure 7.2).

Ball bearings require a low starting torque, whereas plain bearings generally require a high starting torque,[19] as shown in Figure 7.3. This phenomenon is well appreciated in railway prac-tice. It is now accepted that the use of roller bear-ings in trains reduces the starting power needed by several fold, although the running power needed, compared with well-lubricated plain bearings, is similar.[20] Plain bearings are sensitive to load and rotational rate because of the changing character-istics of the lubricant film separating the shaft and bearing. Figure 7.4 shows that under optimum conditions very good performances can be obtained from plain bearings, but the range of coefficient of friction is large for variable conditions of bearing load and speed.

If a plain bearing is not kept well lubricated, the friction can increase many fold. Because of the practical problems involved in carrying out such an operation, and also because of a probably

Figure 7.2
Types of ball bearings.
(a) Annular or radial.
(b) 1893 "Magneto": The
Raleigh had a threaded inner
race.
(c) Cup and cone: From
the diagram it can be seen
how the bearing is self-
aligning and can accom-
modate a bent spindle.

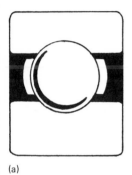  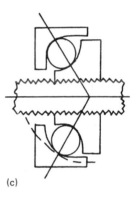

(a)   (b)   (c)

Figure 7.3
Bearing torque for shaft
turning from rest.
Data from reference 19.

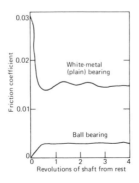

appreciably greater expense and maintenance, it appears desirable to continue the use of ball bearings in bicycles. Some think that plain bearings made from nonmetallic materials could now be used. It has been found that such bearings function in wet conditions, without oil lubrication, desirable features for bicycle bearings. The now well-known Nylon is one of these materials and another is polytetrafluoroethylene or P.T.F.E., a highly corrosion-resistant synthetic chemical polymer.

It is stated in reference 22 that special P.T.F.E. bearings, incorporating metal mixtures in order to overcome certain practical difficulties of ease of seizure (as is experienced with pure P.T.F.E.) have been tested. It appears that the coefficients of friction are 0.10 to 0.16 for suitable loading and design. Table 7.1 gives information about other bearing materials, showing that the minimum coefficient of friction appears to be associated with Nylon 66, a very hard nonmetallic substance. This minimum value, however, is still high, 0.04, which is several times that associated with ball bearings.

Figure 7.4
Friction coefficient of a plain bearing. From G. F. Charnock, *The mechanical transmission of power* (London: Crosby Lockwood, 1932), p. 30. Steel shaft in rigid, ring-oiling, 3-in.-diameter pillow block with gunmetal steps. "Gargoyle Vaculine C" lubricant. This very efficient bearing was probably some four times as easy running as an average plain bearing.

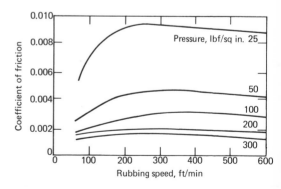

The published facts concerning the performance of nonmetallic bearings, as mentioned above, suggest that if such bearings were used in bicycles, an appreciable increase in resistance to movement would have to be tolerated. It is probable that the power needed to propel a bicycle and rider at 10 mile/h [4.47 m/sec] would be about 1 1/3 times that needed for riding a bicycle (on the level and in still air) fitted with ball bearings. (This estimation assumes that the bearing friction effect is increased tenfold over that associated with ball bearings, which gives an effective resistance of 18 lbf/ton [0.079 newton/kg].) It appears, therefore, that nonmetallic bearings would be suitable only for, say, the machines intended to be ridden by children or certain invalids whose speed of progress it might be desirable to restrict for safety purposes.

Life of bearings

While the life of a plain bearing in a turbine, for instance, is virtually infinite because high-pressure lubrication and high-velocity relative motion combine to prevent metal-to-metal contact, such conditions could not be approached in a bicycle bearing. Short life and high friction must be expected. On the other hand, ball bearings always have a limited life, but the life can be adequate. The time between overhauls of many aircraft turbine engines is well over 20,000 hours, and the bearings are not usually changed. Cup-and-cone ball bearings on bicycles are made of inexpensive steels, inaccurately constructed, and little protected from grit, and can be expected to need replacement after 1,000 hours. However, some specialty manufacturers are supplying wheel hubs incorporating standard automobile-type ball-bearing assemblies to achieve lower friction, longer life, and less maintenance.

The Sturmey Archer type of hub gear is an exception to the suggestions that cup-and-cone bearings and plain bearings have short lives in bicycle use. Effective labyrinth dirt seals are used; the balls are enclosed in cages that eliminate ball-to-ball rubbing; and bearings are accurately aligned, so that long lifetimes are usually experienced. Early Sturmey Archer gears (in 1909) incorporated ball bearings in the mounting of the pinion gears, which was claimed to eliminate 60 percent of the friction. But the bearing loads on these pinion mountings are extremely low, and plain bearings (hardened-steel pins) were substituted without comment later in 1909 and appear to give an acceptable life. The actual, rather than the claimed, effect on the gear efficiency of substituting these plain bearings for ball bearings is not known.

Variable gears

The power loss in an enclosed, lubricated, hub gear arises from the rolling and rubbing of the teeth of mating gearwheels, the friction in their supporting plain bearings, and the squeezing of the several oil films (Figure 7.5). General engineering experience suggests that about 2 percent power loss occurs for each set of mating wheels and 2 percent for a plain bearing. Hence it is probable that a hub gear working with a set of planetary gearwheels on plain bearings could lose, say, 4 to 6 percent of the input power in friction at high power levels and a higher percentage at lower power levels. On the other hand, such a hub when working in direct drive would lose a negligible amount of power.[23] (See Figure 7.6.)

The power loss in a derailleur gear is caused by friction on the hub cogs arising from the sideways rubbing of a chain in a misaligned position; by the added flexing required of the chain in passing around the jockey and tensioner pulleys; and by the friction in the jockey and tensioner pulley bearings, which rotate at relatively high speed. The design conditions and the grit introduced because of the exposed running conditions of such gears are so variable that a dogmatic estimate of power loss is impossible, but for clean, well-lubricated conditions the losses are likely to be about 5 percent.

It is worth noting in connection with the above that the small infinitely variable gears avail-

Figure 7.5
Exploded view of Sturmey-
Archer five-speed hub gear.
Courtesy of Raleigh
Industries Inc.

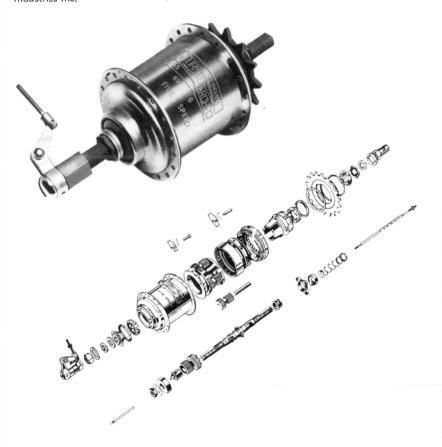

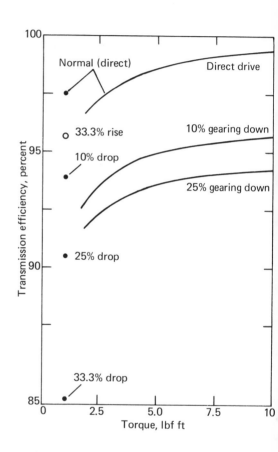

Figure 7.6
Efficiency of hub gears.
Curves are from data in
reference 23. Points are
experimentally determined
by Whitt.

able for general engineering purposes needing a
power output of about 1 hp [746 watts], and
therefore appropriate for bicycle usage, can be
very inefficient. A power loss of 24 percent is

quoted in one advertisement for such gears, although the cost of the gear is many times that of a bicycle hub gear.

The renewed interest in bicycling has brought about a reawakening of inventors, particularly, perhaps, in the variable-gear field. Three designs aim at eliminating chain misalignment and at least one of the tensioner cogs needed on derailleur gears by changing the effective size of either the chainwheel or the wheel cog.

The Tokheim gear (Figure 7.7) has several sets of teeth incorporated in the free-wheel assembly. Each set functions as a cog even though there is not a full set of teeth. Each set can be moved into the plane of the chain. The effective diameters which can be used for the sets of teeth must be such as to leave a clearance between them, so that the choice of gear ratios is limited.

Hagen International Inc. produced a chainwheel with an infinitely variable diameter, within certain limits. With a finite pitch between teeth, chainwheels with an integral number of teeth can vary in size by a minimum of one tooth at a time. Hagen solved this problem by having the chainwheel teeth provided with six cogs or sprockets, which can be adjusted inward or outward in six radial slots (Figure 7.8). The sprockets are mounted on one-way clutches, or free wheels, to permit engagement of the chain in any radial position, while giving a positive-drive capability.

The senior author (FRW) has made a virtue of the varying velocity ratio which these gears give by constructing a chainwheel which is split across a diameter. The two halves are capable of being moved apart by steps to give an effectively oval chainwheel, the ovality increasing as the gear ratio increases (Figure 7.9).

The transmission efficiency of these three gears would be expected to be higher—say 97 percent—than either the hub gear or the true derailleur gear. A chain that connects chainwheel and rear sprocket without tensioners can have a transmission efficiency of 98.5 percent, as stated earlier.

Figure 7.7
Cutaway of Tokheim
transmission showing interaction of Speedisc and
chain. Courtesy of
Tokheim Corporation.

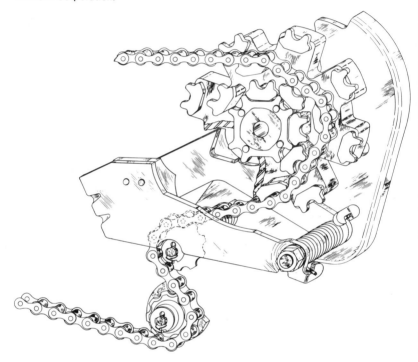

Figure 7.8
Hagen all-speed variable-
diameter chainwheel.
Courtesy of Hagen
International Inc.

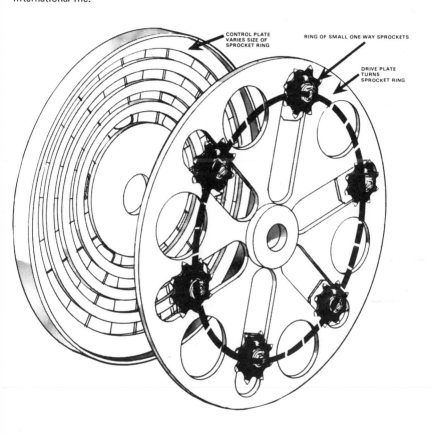

Figure 7.9
Whitt expanding oval
chainwheel gear.

**References
Chapter 7**

1. *Kempe's engineers year book,* vol. II (London: Morgan Brothers, 1962), p. 316.

2. Ibid., p. 128.

3. G. F. Charnock, *The mechanical transmission of power* (London: Crosby Lockwood, 1953).

4. J. Y. Harrison et al., "Maximizing human power output by suitable selection of motion cycle and load," *Human Factors,* vol. 12, no. 3, 1970, pp. 315-329.

5. See references 1 and 3 above.

6. R. F. Kay, *The theory of machines* (London: Edward Arnold, 1952), p. 278.

7. A. Sharp, *Bicycles and tricycles* (London: Longmans, Green and Company, 1896), pp. 396-433.

8. British Standards Institution Publication, B.S228: 1954.

9. American Standards Institution Association publication, specification 1329.

10. C. Bourlet, *La bicyclette, sa construction et sa forme,* (Paris: Gauthier-Villars, 1889), pp. 85-97.

11. See reference 3 above.

12. See reference 6 above.

13. See references 8 and 9 above.

14. See reference 7 above.

15. See reference 1 above.

16. J. L. Koffman, "Tractive resistance of rolling stock," *Railway Gazette* (London),1964, pp. 899-902.

17. Wilhelm von Döbeln, "A simple bicycle ergometer," *Journal of Applied Physiology,* vol. 7, 1954, pp. 222-229.

18. L. Levinson, *Fundamentals of engineering mechanics,* edited by J. Klein, (Moscow: Foreign Languages Publishing House, 1968).

19. Sir Richard Glazebrook, editor, *A dictionary of applied physics* (London: Macmillan and Company, 1922).

20. See also reference 1 above.

21. See reference 19 above, p. 375.

22. See reference 1 above.

23. A. Thom, P. G. Lund, and J. D. Todd, "Efficiency of three-speed bicycle gears," *Engineering* (London), vol. 180, 2 July 1956, pp. 78-79.

Additional recommended reading

Bowden, F. P. and Tabor, D. *Friction and lubrication* (London: Methuen and Company, 1956).

Caunter, C. F. "Cycles - a historical review", Science Museum, reprint series, London, 1972 (for a good review of various types of gearing).

Swann, D. *The life and times of Charley Barden* (Leicester: Wunlap Publications, 1965), p. 58.

Braking of Bicycles

The friction of dry solid substances

Experiments have shown that when two surfaces are pressed together with a force F, there is a limiting value R of the frictional resistance to motion. This limiting value of R is a definite fraction of F, and the fraction or ratio R/F is called the coefficient of friction, μ. Therefore, $R = \mu F$. For dry surfaces, μ is affected little by the area of the surfaces in contact or the magnitude of F.

When surfaces start to move relative to one another, the coefficient of friction falls in value and is dependent upon the speed of movement of one surface past the other. For steel wheels on steel rails, the coefficient of friction can be 0.25 when stationary and 0.145 at a relative velocity of 40 mile/h [17.9 m/sec]. Polishing of the surfaces lowers the coefficient of friction (one cause of brake "fade"), as does wetting.

Coefficients of metal-to-metal dry friction are about 0.2 to 0.4 (down to 0.08 when lubricated); leather-to-metal 0.3 to 0.5. All these are for stationary conditions and decrease with movement.

Brake-lining materials against cast iron or steel have a coefficient of friction of about 0.7, and this value decreases less with movement than for other materials.

Bicycle brakes

Two places where solid-surface friction occurs must be considered in normal bicycle braking: the brake surfaces and the road-to-wheel contact. ("Normal" excludes track bicycles which have no brakes as such: the rider can retard the machine by resisting the motion of the pedals, the rear cog being fixed to the wheel hub without a so-called "freewheel" being used).

Five types of brakes have been fitted to regular bicycles for ordinary road use.

The plunger brake is used on some present-day

children's bicycles and tricycles and was used on early bicycles such as the old ordinary or penny-farthing, and on pneumatic-tired "safeties" up to about 1900 (Figure 8.1). Pulling a lever on the handlebars presses a metal shoe (sometimes rubber-faced) on to the outer surface of the tire. These were and are used on solid and pneumatic tires; the performance is affected by the amount of grit taken up by the tire which fortunately increases braking effectiveness and wears the metal shoe rather than the tire. Such brakes are very poor in wet weather because the tire is being continuously wetted.

The internal-expanding hub brake is similar to the hub brakes of motorcycles and cars, but is less resistant to water, and therefore variable in performance in wet weather. Hub brakes used to be popular for the medium-weight "roadster" type of machine in the thirties, but they have now gone out of favor.

The back-pedaling or "coaster" hub brake brings multiple disks or cones together when the crank rotation is reversed (Figure 8.2). These brakes operate in oil and are entirely unaffected by weather conditions. They are very effective on the rear wheel only: they cannot be fitted to the front wheel because the actuating force required is too great to be applied by hand.

The disk brake has recently been introduced for bicycles in the United States and Japan. At present it is used for the rear wheel only and is cable operated from normal hand levers (Figure 8.3). The effective braking diameter is at less than half the wheel diameter, requiring a high braking force but keeping the surfaces away from the wheel spray in wet weather. These brakes are reputed to be effective in wet and dry weather.

The rim brake is the most popular type: a pad of rubber-composition material is forced against the inner surfaces or the side surfaces of the wheel rims, front and rear. Because the braking torque does not have to be transmitted through the hub and spokes, as for the preceding three types, and

Figure 8.1
Plunger brake on Thomas
Humber's safety bicycle.
Reproduced with permission
from Nottingham Castle
museum.

Figure 8.2
Exploded view of Bendix
back-pedaling hub brake.
Courtesy of Bendix
Corporation, Power and
Engine Components
Group, Elmira, N.Y.

Figure 8.3
Rear-wheel disk brake.
Courtesy of Shimano
American Corporation.

because the braking force is applied at a large
radius, these brakes are the lightest types in themselves and result in the lightest bicycle design. Rim
brakes are, however, very sensitive to water—the
coefficient of friction with regular combinations
of brake blocks and wheel materials has been
found to fall when wet to a tenth of the dry value[1]—
and to rim damage. The composition blocks wear
rapidly and the brakes therefore need continual
adjustment, and block replacement in the order
of 2000 miles [3,218 km]. (Automobile brakes
with heavier duty last around 50,000 miles [80,467
km] before the brake shoes require replacement.)

All present types of brakes have, therefore,
serious disadvantages.

Let us examine the duty required of braking
surfaces for bicycle rim brakes in relation to those
for cars.

Duty of brake surfaces   The brakes for modern motor vehicles can be designed by allowing a certain horsepower—6 to 10—

to be absorbed per square inch [6.94-11.56 $\times 10^6$ watts/sq m] of braking surface for drum brakes.[2] The power to be absorbed depends upon the speed and mass of the vehicle and also on the time in which it is desired to stop.

For a typical bicycle of 30 lbm [13.6 kg] and rider of 170 lbm [77.1 kg], let us determine the power loading at the brake blocks (assumed to have a total area of 4 sq in. [2,581 mm^2]) if a retardation of $-0.5g$ (half gravitational acceleration) from 20 mile/h [88/3 ft/sec or 8.94 m/sec] is required.

Time t for retardation is given by

$$v_2 = v_1 + a\,t,$$

where $v_2 = 0$ and v_1 is the initial velocity. Therefore $v_1 = -a\,t$ and so

$$t = -\frac{v_1}{a} = -\frac{(88/3)\ \text{ft/sec}}{-0.5 \times 32.2\ \text{ft/sec}^2} = 1.822\ \text{sec.}$$

The stopping distance is

$$S = \frac{v_1 + v_2}{2}\,t = \frac{88}{3}\frac{1.822}{2} = 26.7\ \text{ft [8.14 m].}$$

The initial kinetic energy is

$$KE = \frac{mv^2}{2g_c} = \frac{200\ \text{lbm}}{2 \times 32.2\ \text{lbm ft/lbf sec}^2}\left(\frac{88}{3}\ \frac{\text{ft}}{\text{sec}}\right)^2$$
$$= 2,672\ \text{ft lbf [3,627 joule].}$$

The power dissipation falls from a peak at initial application of the brakes to zero when the bicycle comes to rest. For determining brake duty—largely a function of surface heating—the mean power dissipation, KE/t, is required:

Mean power dissipation

$$= \frac{2,672\ \text{ft lbf}}{1.822\ \text{sec} \times 550\ \text{(ft. lbf/sec)/hp}}$$

$$= 2.67\ \text{hp [1991 watts].}$$

Power absorbed per unit of brake-block area $= \dfrac{2.67\ \text{hp}}{4\ \text{sq in.}}$

$= 0.667$ hp/sq in. $[0.771 \times 10^6$ watt/m$^2]$.

This is less than one-tenth of the average loading allowed in automobile-brake practice. Therefore the surface area is more than adequate.

The adequacy of the braking surface fitted to a vehicle is, of course, only one factor in determining the distance in which the vehicle can be stopped. It is necessary in addition to be able to apply an adequate force to the brake system. Bicycle brakes are often deficient in this respect, especially in wet weather when the coefficient of friction is greatly reduced, and especially for the front wheel, where most of the braking capacity is available. Bicycle brakes have not yet been fitted with even a simple type of "servo" system, used for many years on motor vehicles to divert some of the retardation force into braking force.*

Friction between tire and road

If we assume that an an appropriate force can be applied to the brakes and the blocks have been proportioned so that the blocks or linings do not "fade" on account of heating, the stopping capacity of the brakes depends directly upon the "grip" (or coefficient of friction) of the tires on the road. For pneumatic-tired vehicles, this grip varies from 0.8 to 0.1 times the force between tire and road, according to whether the surface is dry concrete or wet ice.

Longitudinal stability during braking

The weight of the bicycle and rider does not divide itself equally between the two wheels, particularly during strong braking. To determine whether or not the braking reaction is important, let us estimate the changes in wheel reactions for the typical bicycle and rider above, braking at half the acceleration of gravity.

If the wheelbase is 42 in. [1.067 m] and the center of gravity of the rider and machine is 17 in. [0.432 m] in front of the rear-wheel center and 45 in. [1.143 m] above the ground (Figure 8.4),

---

*The 1974 Paris bicycle show included a servo-action brake.

Figure 8.4
Assumed configuration
for braking calculations.

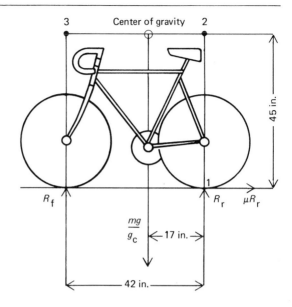

the front-wheel reaction R_f when stationary or
when riding at constant speed is given by

$$R_f \times 42 \text{ in.} = 200 \text{ lbf} \times 17 \text{ in.},$$

where the reaction has been calculated around
point 1 in the figure. Therefore $R_f = 81$ lbf [360
newton]; $R_r = 200 - 81 = 119$ lbf [529 newton].
During the 0.5g braking, a total braking force of
100 lbf [444 · 8 newton] (0.5 $\times$ total weight) acts
along the road surface. The front-wheel reaction
R_f around point 2 in the figure is now

$$R_f \times 42 \text{ in.} = 200 \text{ lbf} \times 17 + 100 \text{ lbf} \times 45 \text{ in.},$$

$$R_f = 81 \text{ lbf} + 107.1 \text{ lbf} = 188.1 \text{ lbf} [837 \text{ newton}];$$

by subtraction:

$$R_r = 11.9 \text{ lbf} [53 \text{ newton}].$$

So the rear wheel is in only light contact with
the ground. Only a slight pressure on the rear brake
will cause the rear wheel to lock and skid. The

front brake has to provide over 90 percent of the total retarding force at a deceleration of 0.5*g* even if the tire-to-road coefficient of friction were at the high level of 0.8. Therefore brakes which operate on the rear wheel only, however reliable and effective in themselves, are wholly insufficient to take care of emergencies.

Another conclusion from this calculation is that a deceleration of 0.5*g* (16.1 ft/sec^2 [4.91 m/sec^2]) is almost the maximum which can be risked by a crouched rider on level ground before he goes over the handlebars.* Tandem riders and car drivers do not have this limitation; if their brakes are adequate they can theoretically brake to the maximum limit of tire-to-road adhesion. If the tire-to-road coefficient of friction is 0.8 then they are theoretically capable of a deceleration of 0.8*g*, which is 60 percent greater than that of a bicyclist with the best possible brakes. For this reason—and many others—bicyclists should never "tail gate" motor vehicles.

Minimum braking distances for stable vehicles

If it is assumed that the slowing effect of air resistance is negligible, a relatively simple formula can be used to estimate the minimum stopping distance of a vehicle fitted with adequate braking capacity, and having the center of gravity sufficiently low or rearward in relation to the wheelbase for there to be no danger of the rear wheels lifting:[3]

$$\text{distance (ft)} = \frac{[\text{initial speed (mile/h)}]^2}{30\left(\begin{matrix}\text{coefficient} + \text{coefficient of} \\ \text{of adhesion} \quad \text{rolling resistance}\end{matrix}\right)}.$$

Table 8.1 gives typical values for the coefficients and Table 8.2 gives calculations for various speeds of a pneumatic-tired vehicle and a railway train. In practice, greater distances are needed for braking than those based on the formula along with an assumption of good grip of tire on its track. The

---

*The deceleration at which this occurs—when the rear-wheel reaction is zero—is about 0.56*g* (17.9 ft/sec^2 [5.45 m/sec^2]).

Table 8.1. Coefficients of adhesion and rolling resistance (motor car).

Surface	Coefficient of adhesion	Coefficient of rolling
Concrete or asphalt (dry)	0.8 - 0.9	0.014
Concrete or asphalt (wet)	0.4 - 0.7	0.014
Gravel, rolled	0.6 - 0.7	0.02
Sand, loose	0.3 - 0.4	0.14 - 0.3
Ice	0.1 - 0.2	0.014

Sources:
Reference 2, p. 321.
G. M. Carr and M. J. Ross, "The MIRA single-wheel rolling resistance trailers," Motor Industries Research Association, Nuneaton, Warwickshire, England, 1966.

Table 8.2. Stopping distances for bicycles, cars, and trains.

Speed, mile/h	Stopping distance, pneumatic tires, ft			Railway train, practical, ft
	Formula	Safety code cycle	Safety code car	
8	2.5	3		40
10	4			60
12	5.7	8		80
16	10	16		120
20	16	24	20	160
30	36		45	260
40	64		80	510
50	100		125	850
60	145		185	1300

Note: The adhesion coefficient used for calculated stopping distances is 0.85. The other distances for pneumatic tires are quoted from Road Safety Codes. All values are for stopping on dry concrete. Practical values for railway trains are included for comparative purposes.

railway figures indicate that if an adhesion coefficient of 0.1 is assumed, the formula gives braking distances of about half those normally found in practice.[4]

Table 8.2 includes distances quoted in British road-safety codes[5] for best performance of pneumatic-tired vehicles. These are also about twice those calculated from the formula (with an assumed adhesion coefficient of an achievable magnitude under very good circumstances). The road-safety-code performance figures have been well checked by the Road Research Laboratory, UK, the 1963 report of which gives details of measurements carried out on "pedal cycles" of various types as well as many types of motor vehicles.[6] The braking distances listed for bicycles confirm the calculations made above, where it was found that a little better than 26 ft [8.14 m] was possible for stopping from 20 mile/h [8.94 m/sec] without overturning. If the rider sat well back over the rear wheel he would be able to shorten the distance a little further. However, evidence obtained from spot checking indicates that the average motor vehicle on the road needs about twice the quoted code distances for braking under specified conditions,[7] and it may be assumed that the same "service factor" applies to bicycles.

Braking on the rear wheel only

Let us see what braking distance we may expect if the same rider and bicycle studied earlier, starting from 20 mile/h [8.94 m/sec], brake with the rear brake only to the limit of tire adhesion. We assume that the rear brake is strong enough to lock the wheel if desired, and that the coefficient of friction μ between the tire and the road surface is 0.8. Then the maximum retarding force is $0.8 \times R_r$, where R_r is the perpendicular reaction force at the rear wheel. This rear-wheel reaction force R_r is somewhat less than the value during steady level riding or when stationary because the deceleration results in more reaction being taken by the front wheel. Let us take the moments of forces (torques) about point 3 in Figure 8.4.

Under the assumed static conditions the machine is in equilibrium:

$$R_r \times 42 \text{ in.} + \mu R_r \times 45 \text{ in.} = \frac{mg}{g_c} \times 25 \text{ in.}$$

$$R_r = \frac{200 \text{ lbm} \times 32.2 \text{ ft/sec}^2 \times 25 \text{ in.}}{32.2 \text{ lbm ft/lbf sec}^2 \, (42 \text{ in.} + 0.8 \times 45 \text{ in.})}$$

$$= \frac{200 \times 25 \text{ lbf}}{(42 + 36)} = \frac{5{,}000 \text{ lbf}}{78} = 64.1 \text{ lbf } [285.3 \text{ newtons}],$$

where we have assumed the sea-level value, 32.2 ft per $\sec^2$, for g, m = 200 lbm [90.72 kg], and $\mu = 0.8$.

Then the deceleration as a ratio of gravitational acceleration is given from Newton's law,

$$F = \frac{ma}{g_c},$$

$$a = \frac{Fg_c}{m} = \frac{-\mu R_r g_c}{m},$$

$$\frac{a}{g} = \frac{-\mu R_r g_c}{m \quad g}$$

$$= \frac{-0.8 \times 64.1 \text{ lbf}}{200 \text{ lbm}} \quad \frac{32.2 \text{ lbm ft/lbf sec}^2}{32.2 \text{ ft/sec}^2}$$

$$= -0.256;$$

$$a = -0.256 \, g.$$

So the retardation is just about half the value at which, using the front brake to the maximum, the rider would go over the handlebars. (We know that this value is a little over $-0.5 \, g$).

The time taken for this deceleration is given as before by

$$v_1 = -at,$$

$$t = \frac{-(88/3) \text{ ft/sec}}{-0.256 \times 32.2 \text{ ft/sec}^2} = 3.56 \text{ sec.}$$

The stopping distance is given by

$$S = \frac{v_1 + v_2}{2} \, t = \frac{88}{3} \frac{3.56}{2} = 52.2 \text{ ft (15.91 m).}$$

Therefore the stopping distance is about twice that for reasonably safe front-wheel braking. In practice, a longer stopping distance is likely because a deceleration level sufficiently below the limit where skidding starts would be chosen.

Wet-weather braking

Wet conditions affect both road adhesion and, with bicycle rim brakes, the brake grip on the rim. Braking distances for bicycles are approximately quadrupled in wet weather.[8] Cars are generally fitted with weather-proof drum brakes and are not affected by wet weather to anything like the extent that are bicycles.

Experiments using laboratory equipment to simulate wet-weather braking of a bicycle wheel have also been carried out.[9] The most significant findings from these two sets of experiments were as follows.

For brake blocks of normal size and composition running on a regular 26-in. [equivalent to 650 mm] diameter plated steel wheel, tests at Massachusetts Institute of Technology showed that the wet coefficient of friction was less than a tenth of the dry value (Figure 8.5).[10] Moreover, the wet wheel would turn an average of 30 times with full brake pressure applied before the coefficient of friction began to rise, and a further 20 turns were necessary before the full dry coefficient of friction was attained (Table 8.3). This recovery was possible only if no water was being added to the brake blocks or rims after brake application, as might occur during actual riding in very wet conditions.

Figure 8.5
Wet versus dry braking-
friction coefficients.
Rim material: steel, nickel-
chromium plated.
Data from reference 1.

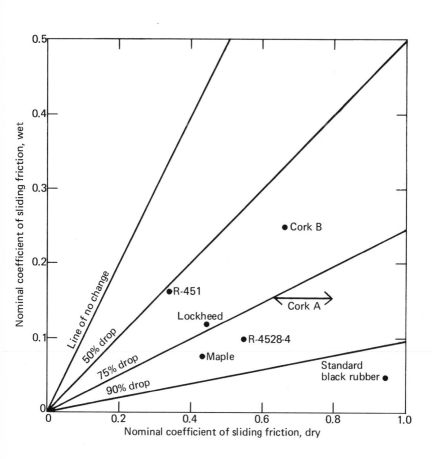

Table 8.3. Test-rig data on wet operation of rim brakes. Run C-2, 19 January 1971; nature of run, wet-dry.

Point	Braking force lbf	Coefficient of friction, μ
1(wet start)	22	0.17
2(prerecovery)	22	0.17
3(recovering)	26	0.20
4(recovering)	31	0.24
5(recovering)	35	0.27
6(recovering)	39	0.30
7(recovered)	44	0.34
Turns of wheel before onset of recovery		30
Turns of wheel during recovery		20
Total turns to recovery		50

Source: Reference 1, page 32.

Table 8.4. MIT Tests on brake block materials.

Run	Friction material	Equiv. speed, mile/h	Nature of run	Average μ_{dry}	Average μ_{wet}	$\dfrac{\mu_{wet}}{\mu_{dry}}$	Turns to recovery	Remarks
C-1	R-451	10	dry	0.33	—	—	—	μ = 0.39 at 120°F
C-2	R-451	10	wet-dry	0.34	0.17	0.50	50	
C-3	B. rubber	10	wet-dry	0.95	0.05	0.05	55	Erratic recovery
C-4	R-4528-4	10	wet-dry	0.55	0.10	0.18	54	
C-5	Maple	10	wet-dry	0.44	0.09	0.20	42	μ_{max} = 0.56
C-6	Lockheed	10	wet-dry	0.45	0.12	0.27	25	during rec'y
C-7	R-451	10	wet-dry	0.34	0.17	0.50	53	
E-1	Cork A[a]	10	dry	0.63	—	0.42	—	
E-2	Cork A	10	wet	—	0.26			
E-3	Cork A	10	dry	0.79	—	0.24	—	
E-4	Cork A	10	wet	—	0.19			
F-1	Cork B[b]	10	dry	0.67	—	0.28		
F-2	Cork B	10	wet	—	0.19			
F-3	Cork A	10	wet[c]	—	0.16	—	—	
F-4	Cork B	10	wet[c]	—	0.25	—	—	
F-5	R-451	10	dry	0.43	—	—	—	
F-6	R-451	10	wet-dry	0.37	0.17	0.46	70	

Source: Reference 1, p. 34.
[a]Orientation A: Layers parallel to friction face
[b]Orientation B: Layers perpendicular to friction face
[c]After a 48-h soak

A number of different materials were investigated at MIT, and the results are shown on Figure 8.5 and Table 8.4. Although many of the materials are brake materials designated only by numbers, it can be seen that regular bicycle brake blocks ("B-rubber") have the highest dry coefficient and the lowest wet coefficient of friction of all materials tested. Attempts to improve the wet friction by cutting various grooves in the blocks or by using "dimpled" steel rims were unsuccessful.

The Road Research Laboratory found that wet-weather performance can be improved by the use of brake blocks longer than the usual 2 in. [5.1 cm].[11] Softer blocks than are common these days are also desirable, along with more rigidity in the brake mechanism and in the attachment to the frame of the brake itself.

The longer rear-cable mechanism can, because of extra cable friction, decrease the force applied by a rider at the blocks by 20 percent compared with that at the front brake. However, it has been pointed out that the rear brake requires less actuating force than does the front if locking (skidding) is to be avoided. Virtually no present brakes allow adjustment without resort to wrenches through the whole range of brake-block wear, a lack which leads to extremely dangerous conditions in bicycles ridden by the less mechanically able persons.

Adhesion of tires

It has been found that even when surfaces roll upon one another, a certain amount of "slipping" takes place which, in turn, leads to frictional losses. This phenomenon is rooted in the fact that the surfaces, however "hard," do create cavities at the points of contact, and this leads to alternate compression and expansion of the materials at these points and, as a consequence, expenditure of energy.[12] With soft surfaces, of course, the effects are pronounced but are well worth putting up with where vehicle tires are concerned because of the comfortable riding produced.

Although efficiently functioning tread patterns are essential for the good grip of motorcar tires on the road under high-speed wet conditions, it appears that at the low speeds used by bicycle riders bicycle-tire requirements are not so stringent. Data given in some tests suggest that no appreciable variation in the grip of a tire on the road under wet conditions could be expected from any design alteration.[13] At low speeds of under 20 mile/h [8.94 m/sec] nearly smooth patterns of tread should suffice. Indeed this prediction is verified by the but slightly corrugated tire surfaces of racing tires used over many years of cycle-tire manufacture.

Braking by means of back-pedaling

As stated earlier, track bicycles have no separate brakes, and riders slow down by trying to "back pedal" on the "fixed-wheel" (the rear wheel is not fitted with any device that allows free wheeling). The idea that the rider should perform work to destroy energy has intrigued many people since the early days of bicycling. Horse-drawn vehicles have braked in this way for thousands of years, and men running down stairs and steep slopes experience a similar muscular action.

Much discussion was devoted in the past to arguments about muscular actions concerned with forward and backward pedaling by comparison. Sharp[14] concluded that muscle physiology played an equal part with mechanical motion. He devised the interesting chart, Figure 8.6, in the course of his writings on the subject. The passage of time has proved his surmise correct in that research workers have shown that for a given oxygen consumption a pedaler can resist power supplied by an animate or inanimate prime mover to a greater efficiency than he performs with ordinary pedaling.[1] A now classic experiment of a normal ergometer pedaler being resisted by a pedaler in reverse vividly demonstrated this difference in energy cost for forward pushing as distinct from resisting. The basic physiological reasons for this matter involving muscle-action theory are still being debated in the literature under the heading of "negative work," sometimes called "eccentric work."

Braking of bicycles

Figure 8.6
Power expended in back
pedaling.
The dashed lines are
resistance curves and
represent rolling plus
aerodynamic drag.
The solid lines are power
curves.
g is the gradient expressed
as percentage/100 (for
example, 0.12 is 1 in 8.5).
Intercepts of power curves
with the horizontal axis
show terminal downhill
speeds for each gradient.
Between these velocities
and zero velocity the
"negative" power that has
to be exerted in back
pedaling goes through a
maximum.
From reference 14.

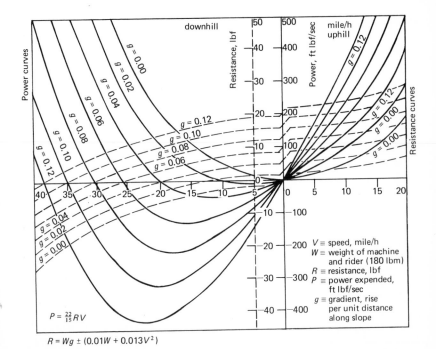

References
Chapter 8

1. Brian D. Hanson, "Wet-weather-effective bicycle rim brake: an exercise in product-development," MS thesis, mechanical-engineering department, Massachusetts Institute of Technology, Cambridge, Massachusetts, June 1971.

2. *Kempe's engineer's year book,* vol. II (London: Morgan Brothers, 1962), pp. 320 and 353.

3. Ibid.

4. Ibid.

5. "Safe cycling," Her Majesty's Stationary Office, London, February 1957.

6. "Research on road safety," Her Majesty's Stationery Office, London, 1963.

7. Ibid.

8. Ibid.

9. See references 1 and 6 above.

10. See reference 1 above.

11. See reference 6 above.

12. Sir Richard Glazebrook, editor, *A dictionary of applied physics* (London: Macmillan and Company, 1922).

13. See reference 6 above.

14. A. Sharp, "Back-pedalling and muscular action," *CTC Gazette,* 1899, pp. 500-501.

15. B. C. Abbott, Brenda Bigland, and J. M. Ritchie, "The physiological cost of negative work," *Journal of physiology,* vol. 117, 1952, pp. 380-390.

16. H. B. Falls, *Exercise physiology,* (New York: Academic Press, 1968), pp. 292-294.

Additional recommended reading

Kay, R. F. *The theory of machines* (London: Edward Arnold, 1952).

Bicycle balancing and steering

The earliest man-propelled road vehicles appeared as three or four-wheelers closely resembling the lighter horse-drawn carriages of the period of 1700 onwards. These were inherently stable in that they normally needed no balancing on the part of the occupant. They were, like their animal-drawn counterparts, fitted with a steerable front wheel or pair of wheels.

In contrast, the first single-tracked two-wheeled man-propelled vehicles were without a means of steering. The two wheels were fixed rigidly in one plane. An illustration from an early book on cycling by H. H. Griffin[1] shows an outline of such a vehicle, called then a hobby horse according to the title of the poem attached (Figure 9.1). This 18th-century vehicle did not survive into the 19th century; it was ousted by steerable two-wheelers devised by various inventors such as Baron von Drais de Sauerbrun, Denis Johnson, and others and called by various names. These are all now covered by the earlier name of "hobby horse." These machines, as is described in any of the books on the bicycle such as that by Griffin, lasted only during the early years of the 19th century, a time during which they were much favored as novelties by the Regency Dandies.

Analyses of bicycle stability

In the early days of the bicycle, mathematicians were intrigued with the theory of its unique type of motion. Two early analysts were F. J. W. Whipple[2] and G. T. McGaw,[3] and they were followed later by such renowned figures as Timoshenko and Young.[4] None of their theories were widely accepted, particularly among bicyclists, because they failed to explain commonly experienced aspects of bicycling. Does a bicyclist balance by steering into the fall? Is caster action necessary for balance? Are gyroscopic effects in the front wheel important?

Figure 9.1
Hobby horse. Reproduced
from reference 1.

Ye Hobby-Horse.

" Though some perhaps will me despise,
    Others my charms will highly prize,
    (Yet, nevertheless, think themselves wise.)
    Sometimes, 'tis true, I am a toy,
    Contrived to please some active boy ;
    But I amuse each Jack O'Dandy,
    E'en great men sometimes have me handy !
    Who, when on me they get astride,
    Think that on Pegasus they ride."

County Magazine, 1787.

The theories did not answer these questions. Experiments have. A research scientist, David E. H. Jones, set out to build an "unridable bicycle."[5] He reversed the front fork to nullify caster action. He fitted a counter-rotating wheel on the front fork to nullify gyroscopic effects. He drastically changed other aspects of the steering geometry. But he could still balance and steer quite easily. Only when he locked the front-wheel steering and attempted to steer with the rear wheel did he produce a machine that defeated his efforts to remain balanced.

Jones disproved some hypotheses about balancing and steering, but he was not able to substitute a simple theory of his own. He concluded that the subject was far more complex than mathematicians had first assumed.

A simple approach to mathematical modeling would simulate the rider and machine as a single rigid mass, with two wheels that faithfully direct motion along the plane of the wheels. An actual machine-rider combination differs from this simple picture in at least the following respects.

1. Tire slip: when there is a side force on the wheels, such as when there is a side wind, or when the bicycle is being ridden along a sloping surface, or when a curved path is being followed, the tires "slip" in the direction of the side force. The angle of slip depends on the ratio between the side force and the normal force, on the angle between the plane of the wheel and the ground, on the tire pressure, and on the tire construction. Typical graphical relations for the slip angle are shown in Figure 9.2.

2. Steering angle and trail: early bicycles had a steering angle of ninety degrees and no trail, as shown in Figure 9.3, while today steering angles are about seventy degrees with the wheel-road contact point trailing behind the extrapolation of the steering line (Figure 9.4). This complicated geometrical arrangement produces a self-restoring moment when the wheel is turned, but this moment is affected by bicycle angle, path curvature, and other factors.

Figure 9.2
Typical bicycle-tire slip
angles for various inclinations. Reproduced from
reference 7.

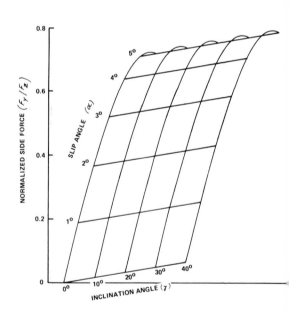

Figure 9.3
Straight forks used in
early bicycles.
(a) French Célérifère,
1816. Reproduced from
reference 1.
(b) English Dandy Horse,
1820. Reproduced from
reference 1.
(c) Boneshaker, 1869.
Reproduced from A. Sharp,
Bicycles and tricycles
(London: Longmans,
Green & Company, 1896),
p. 148.

(a)

(b)

(c)

Figure 9.4
Geometry of the offset
front fork.
ab is tangent to wheel.
ab = *y* = fork offset
be = *z* = trail
y = *z*
This geometry gives no rise
or fall of the frame when
the fork is 90°.
Reproduced from refer-
ence 10.

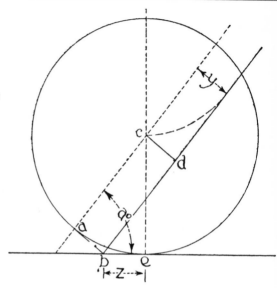

Figure 9.5
Comparison of simulated
and experimental bicycle
responses after a steering
torque disturbance.
Reproduced from
reference 7.

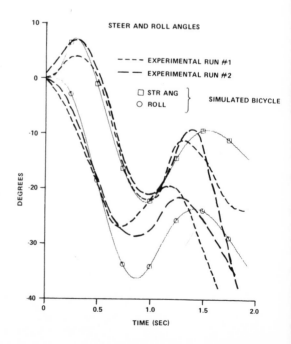

3. Rider steering response: the bicycle rider responds to perceived changes in balance, for instance, by moving the handlebars. Each rider has a different response, and a different delay before initiating the response, thus further complicating the analysis of steering behavior of bicycles plus riders. Human beings are extremely adaptable in their responses, as was shown by Jones.[6]

4. Wheelbase: short-wheelbase bicycles are said to be "responsive" while long-wheelbase bicycles, such as tandems, are "sluggish," for obvious reasons.

5. Bicycle mass: the mass, or weight, of a bicycle, and the point at which this center of mass is located, affects the steering behavior.

6. Rider mass: the mass of the rider, or more particularly the relation of the rider mass to the mass of the machine, and the relative position of the center of mass of the rider, have influence on steering behavior.

7. Wheel moment of inertia: the higher the moment of inertia of the wheel, the higher is the gyroscopic torque produced when the plane of the wheel is turned.

8. Bicycle inclination angle: the angle the bicycle makes with the road significantly affects the steering forces and tire-slip angle (Figure 9.2).

9. Angle of rider: many riders try to hold themselves in the same plane as the bicycle under all conditions, while others may hold their bodies at an angle to the plane of the bicycle, particularly when riding around a curve. In doing so they produce a bicycle inclination angle different from that which would be given if the rider center of mass remained in the plane of the bicycle, and the steering response is changed.

10. Rider - bicycle connection: the bicycle may be ridden with the feet in toe clips, the crotch firmly on an unsprung saddle, and the hands gripping the metal of the handlebars. Conversely, a rider may have a much looser or more flexible connection with a bicycle through a deeply sprung saddle, sponge-filled handlebar grips, and rubber-

tread pedals. Or he may ride with his hands off the
handlebars, or his crotch off the saddle, or his feet
off the pedals. In all these circumstances the re-
ponse of the machine varies.

This list by no means exhaust the components
that contribute to bicycle-riding characteristics—
for instance, the springiness of the bicycle frame,
and the slack and friction in the steering bearing
are of some importance—but these are probably
the most important factors. Many of these factors
are nonlinear. Mathematical analysis is understand-
ably ineffective in such a system. Computer simulation is more appropriate. A simulation by Roland[7]
has been the most comprehensive so far attempted
and has shown considerable success. An example
of the simulation and of angles measured from an
instrumented rider and bicycle is shown in Figure
9.5. (Side-force loading was supplied by firing a
small rocket motor attached to the bicycle frame.
Normally, side forces provided by rocket motors
are hazards not often encountered on American
roads.) The computer was programmed to provide
not only graphical responses but also to illustrate
the rider and bicycle in an elementary form as
shown in Figure 9.6.

The results of Roland's study at Calspan were
that bicycle speed has a more pronounced effect
on stability than do any of the other components
of the system. All configurations examined were
stable and well behaved at 15 mile/h [6.71 m/sec],
while all showed an oscillatory form of instability
at 6 mile/h [2.68 m/sec]. (Obviously bicycles can
be ridden at much lower speeds than 1 mile/h [0.45
m/sec], and the 6 mile/h limit reflected the choice
of the rider-response characteristics.) Wheelbase
was found to be the single parameter having the
greatest effect on stability. The short-wheelbase
configuration was better behaved at low speed
and showed a damped oscillatory response with
only one-half the amplitude of steer correction
which was found to be needed in the long-wheelbase configuration. At 10 miles/h [4.47 m/sec]
the difference between the two configurations be-

Figure 9.6
Computer graphics
rendition of a bicycle
and rider. Reproduced
from reference 7.

came smaller, and at 15 miles/h [6.71 m/sec] it was insignificant. Reducing the steering trail also increased stability, but reducing the total bicycle weight and increasing the height of the center of mass decreased low-speed stability.

This computer simulation confirmed Jones' findings that quite large changes in configuration, with the exception of changes in the wheelbase, have a comparatively small effect on ridability.[8] Most experienced riders would agree with the other findings from the study. This is not to imply that the computer study was not worthwhile. The very fact that the results seem so reasonable and expectable gives confidence that the technique can be applied to new and unsolved problems. For instance, when a sudden front-tire blowout is experienced on a small-wheeled bicycle, the machine frequently becomes unsteerable and the rider can be placed in great danger. This phenomenon is rare on large-wheel machines. Should small wheels be therefore banned? Or are there combinations of steering angle, trail, flat-tire characteristics, and so forth, which would produce a fail-safe system? It is obviously more effective and less expensive to find the answer to this question by operating a computer model than by experimenting with hardware and/or human lives.

Frame and fork design

If present-day mathematics gives little direct guidance as far as bicycle design is concerned, constructors will still have to rely on the long experience now available for setting steering characteristics. Fortunately bicycle frames have become relatively stabilized in design, for reasons other than steering characteristics. All that can now be done with frame design is to alter the head angle and fork offset. The variation, once available, of using large front wheels in safety bicycles is not now acceptable to purchasers. The virtues of the large front wheel, as far as easing steering problems, may have resided in the fact that the bigger wheel was less disturbed by uneven roads than was a small wheel. Nowadays roads are better. Jones

showed that gyroscopic effects are not as important as Victorian advocates of the large front
wheel would have had bicyclists think.

When front-driver bicycles were the vogue the
designer generally put in a straight front fork and
little or no inclination of the head (Figure 9.3).
The feet, of course, could supplement the steering
movement of the handlebar. This action was lost
when rear drivers became the fashion. The fork
offsets and inclinations of the latter-type machines
then became subjects for much debate and experiment in order to get a tolerably ridable bicycle.
Many years of experience have led frame constructors to adopt the combinations of fork offset and
head angle typified by those given in Table 9.1.

One reason for the acceptability of these combinations could be that the turning of the front
fork and wheel, with the machine vertical, gives
neither rise nor fall to the frame. In view of the
observations put forward by Jones about the
effects of inclining the man-machine combination,[9]
the basic steering phenomenon could be much
more complex than just ensuring that there be no
rise or fall of the frame. Figure 9.4 (from Davison[10])
is given to show how the geometry of the fork and
frame head angle can be related. Bourlet gives a
rather complicated formula[11] which assumes a limit
of 2 to 3 cm to the sideways movement of the
head of the bicycle. Bernadet[12] discusses the application of this formula, and the discussion is
continued in reference 13 in which the original
Bourlet relation is requoted.

Table 9.1. Relationship between steering angle and fork offset.

Steering angle a, degrees	Fork offset Y (Figure 9.3) in.	Formula (reference 9)	Formula (reference 10, p. 60)
68	2.59	$Y = R \tan\left(\dfrac{90^\circ - a}{2}\right).$	$\left(\cos a - \dfrac{Y}{R}\right)^3 \left(\cos a + \dfrac{Y}{R}\right) - \dfrac{e^2}{R^2} \cos^2 a\,(1 - \cos^2 a) = 0,$
70	2.36	where R is radius of wheel	where e is the sideways movement of the frame caused by fork turning. This is recommended to be very small with 0.8 in. [2 cm] considered reasonable
72	2.12	No rise or fall of frame occurs when fork is turned	
75	1.77	(27-in. wheel)	With an e value of 0.8 in. [2 cm] the steering angle to satisfy the above equation is 75° and the fork offset is about 1.7 in. [4.5 cm] (27.5 in.-wheel)

Note: In practice great accuracy in estimating the fork offset for a given steering angle is not justified because the sinking of the pneumatic-tired wheel alters the radius R of the wheel involved in the formulas above. On this account it appears that recommendations based on the formula of A. C. Davison (reference 9) and that for 75° angle frame using the formula of C. Bourlet (reference 10) are very similar.

References
Chapter 9

1. H. H. Griffin, *Cycles and cycling* (London: George Bell and Son, 1890).

2. F. J. W. Whipple, "The stability of the motion of a bicycle," *Quarterly Journal of Pure and Applied Mathematics,* vol. 30, 1899, pp. 312-348.

3. G. T. McGaw, *Engineer* (London), vol. 30, 2 December 1898.

4. S. Timoshenko and D. H. Young, *Advanced dynamics,* (New York: McGraw-Hill Book Company, 1948), p. 239.

5. D. E. H. Jones, "The stability of the bicycle," *Physic. Today,* April 1970, pp. 34-40.

6. Ibid.

7. R. Douglas Roland, Jr., "Computer simulation of bicycle dynamics," Calspan Corporation, Buffalo, N. Y., ASME paper, fall 1973.

8. See reference 5 above.

9. See reference 5 above.

10. A. C. Davison, "Upright frames and steering," *Cycling,* 3 July 1935, pp. 16, 20.

11. C. Bourlet, *La bicyclette, sa construction et sa forme,* (Paris: Gauthier-Villars, 1899), p. 60.

12. E. Bernadet, "L'étude de la direction," *Le Cycliste,* September-October 1962, p. 228.

13. Cyclotechnie, "L'étude de la direction," *Le Cycliste,* November-January 1973.

Additional recommended reading

Bower, George S., "Steering and stability of single-track vehicles," *The Automobile Engineer,* vol. V, 1915, pp. 280-283.

Collins, Robert Neil, "A mathematical analysis of the stability of two-wheeled vehicles," Ph.D. thesis, University of Wisconsin, 1963.

C. T. C. Gazette. February 1899, p. 73.

Delong, Fred. "Bicycle stability," *Bicycling,* May 1972, pp. 12, 13, and 45.

Dohring, E. "Stability of Single-Track Vehicles," *Forschung Ing. Wes.,* vol. 21, no. 2, 1955, pp. 50-62 (translation by Cornell Aeronautical Laboratory, Inc., 1957).

Dohring, E. "Steering wobble in single-track vehicles," *Automobil technische Zeitschrift,* vol. 58, no. 10, pp. 282-286 (M. I. R. A. Translation No. 62167).

Fu, Hiroyasu. "Fundamental characteristics of single-track vehicles in steady turning," *Bulletin of Japan Society of Mechanical Engineers,* vol. 9, no. 34, 1965, pp. 284-293.

Kondo, M. "Dynamics of single-track vehicles," Foundation of Bicycle Technology Research, 1962.

Manning, J. R. "The dynamical stability of bicycles," Department of Scientific and Industrial Research, RN/ 1605/JRM, July 1951, Road Research Laboratory, Crowthorne, Berkshlreg, England.

Pearsall, R. H. "The stability of the bicycle," *Proceedings of the Institute of Automobile Engineering,* vol. XVII, 1922, p. 395.

Rice, R. S. and Roland, R. D. "An evaluation of the performance and handling qualities of bicycles," VJ-2888-K, 1970, Calspan Corporation, Buffalo, N.Y. (prepared for the National Commission on Product Safety).

Sharp, R. S. "The stability and control of motorcycles," *Journal of Mechanical Engineering Science,* vol. 13, no. 4, August 1971.

Singh, Digvijai, V. "Advanced concepts of the stability of two-wheeled vehicles, application of mathematical analysis to actual vehicles," Ph.D. thesis, University of Wisconsin, 1964.

Van Lunteran, A. and Stassen, H. G., "Investigations of the characteristics of a human operator stabilizing a bicycle model," Intern. Symposium on Ergonomics in Machine Design, Prague, 1967, p. 27.

Van Lunteren, A. and Stassen, H. G. "On the variance of the bicycle rider's behavior," 6th annual conference on manual control, Wright-Patterson AFB, Ohio, 1970.

Weir, D. M. "Motorcycle handling dynamics and rider control and the effect of design configuration on response and performance," University of California, Los Angeles, 1972.

Wilson-Jones, R. A. "Steering and stability of single-track vehicles," *Proceedings of the Institute of Mechanical Engineers,* Automobile Division, London, 1951-1952.

The makers of early bicycles used "traditional" materials of construction—woods reinforced with metals—the origin of which dates from the earliest vehicles. The shortcomings of this type of construction when applied to a man-propelled vehicle soon became apparent, and tubular-steel construction with bearings which rolled internally, instead of rubbing, appeared in the 1870s. In general there has been no basic change in this attitude toward the basic principle of bicycle construction, although smoother roads, better steels, aluminum alloys, and improved design have resulted in a reduction in bicycle weights to about one-third of that common for early machines.

Almost a century has passed since the design philosophy mentioned above was first established. On this account, and because of the great publicity given to the successful use of more modern man-made materials in certain engineering applications, critics of bicycle construction frequently say that bicycle makers are slow in taking up new ideas and that, if expense were ignored, better (generally meaning lighter and "faster") machines could be made.

Properties of materials of construction

Engineering science has advanced sufficiently for reliance to be placed upon the results of certain standardized tests when used to calculate whether or not a certain material is appropriate for a given structure. Table 10.1 lists the most important characteristics of some materials of construction likely to be contemplated for bicycles. Except for Young's modulus the terms should be self evident. In simple language the Young's modulus figures give a measure of the elasticity, and, hence, the "rigidity." A high value of Young's modulus signifies a stiff material.

Calculations show that in spite of the great

Table 10.1. Properties of materials of construction (typical approximate values).

Material	Tensile strength tonf/in.^{2}	Elongation at failure, percent	Young's modulus lbf/in.^{2} X 10^6	Specific gravity
Aluminum	5 - 9	20 - 30	10	2.5 - 2.6
Duralumin	26	10 - 12.5	10	2.5 - 2.8
Copper	13.4	40	15	8.8 - 9.0
Nickel	38 - 45	20 - 35	20	8.9
Cast iron	8		18	7.0 - 7.2
Wrought iron	25	25	28	7.6 - 7.9
Magnesium alloys	11 - 20	3 - 12	6.5	1.75
Titanium	40		15.8	4.5
Mild steel	28 - 30	16 - 30	30	7.8
High-tensile steels	37 - 49	14 - 22	30	7.8
Stainless steels	50	20	27	7.75
Ash, beech, pine, oak	5 - 7		1.5	0.5 - 0.88
Polymethyl-methacrylate	4 - 5		0.44	1.19
Nylon	4 - 5	80 - 100	0.3	1.14
Glass-fiber-reinforced epoxy	18		3.3	1.5
Glass fiber	22		3.3	1.8

Sources:
Kempe's engineers hand book (London: Morgan Brothers, 1962).
Mechanical world year book, 1967 (Manchester, England: Emmott & Company, 1967), p. 158.

tensile strength per unit weight of the competitors, the high Young's modulus of the steels puts them in the forefront for producing a structure which must have the minimum flexibility per unit weight —a most desirable feature in bicycle construction. The steel structures are also less bulky when structures of given rigidity are compared. A fair example of this is the case of crank-sets in the various aluminum alloys which are always large compared with high-class steel sets.

A feature of the newer metals and plastics when compared with steel is their greater resistance to atmospheric corrosion. The surface treatments necessary to ensure satisfactory service are minor operations compared with the plating or enameling processes inseparable from the use of the steels. On this account the use of the newer materials for the less stressed parts of bicycles has been fairly satisfactory and will no doubt continue to be experimented with in various ways. It is interesting to note that celluloid mudguards were in use in Victorian times, and aluminum structures also appeared and disappeared.

In all the discussion above it has been assumed that cost was not a factor. In the majority of circumstances the costs of the raw materials for manufacture affect the choice. At present, steel is the least expensive material for making a bicycle. When the cost of manufacture is added, it is possible that high-strength plastic may win a place because of the automated production this material allows.

Bearings, chains, and gearwheels in non-metallic materials

Fabricators of machine parts in plastics, in particular, have lately made great advances toward producing competitors to metal parts where silence of running and light weight are important factors. If corrosion resistance matters greatly as, for instance, in chemical plants, nonmetallic parts may have considerable advantages over metal parts.

When applied to most conditions of bicycle usage, plastic components show serious drawbacks compared with corresponding metal parts. Plastic bearings must be made with larger clearances than plain metal bearings, that is, the fit is "sloppier." Nylon 66, a very hard plastic, is the most slippery material from which to make a bearing, but its minimum coefficient of friction of 0.04 shows a fourfold greater resistance to movement compared with a reasonably good ball bearing's performance of 0.01. Several firms now make plastic chains and toothed reinforced rubber belts. All need to be very large, and the complete drive for a bicycle would weigh more than a modern steel chain drive. The chainwheels would likewise be very wide (¾-in.-wide teeth at least) and hence cumbersome. Although gear wheels in nylon are successfully used for small hand-drills, it appears, because of the low strength of the material compared with steel, that a nylon hub gear would be a very bulky object compared with the standard steel hub gear. Such characteristics may be of but little importance for general engineering usage but they are very unpopular in the specialized cycle world.

Great enthusiasm was expressed in an article by a well-known cycle designer, I. Cohen, published in 1955,[1] for the use of a hard plastic, polytetrafluoroethylene (P. T. F. E.) for bearings. However, it has been found that the compressibility of the plastic has caused a great deal of trouble. Plastics of various types have since been used for bearings fitted to children's machines, and some complete parts, such as small pedal frames, have been marketed. It is probable that children are insufficiently discriminating about easy running in their bicycles and tricycles—and are parents not unhappy about the slowing up of their children. Apparently manufacturers have now realized that with machines for adult use there is no doubt that the buyers will not accept plain bearings in plastic or, as has been tried recently for pedals, plain metal bearings. There appears to be a realization that ball bearings are essential for an adult machine to ensure easy running and a reasonably long life without constant adjustment to avoid unsafe "slop." No doubt users of plain bearings fitted to machines in the

1870s to 1890s were glad to see their abandonment, and present-day veteran-cycle enthusiasts owning such machines will endorse the opinion.

Frames in nonmetallic materials

Woods: Bicycle frames of wood have been made and have been ridden with satisfaction at regular intervals since the earliest "hobby-horse" days of about 1800. The Macmillan rear-driver bicycle was introduced in 1839 and was followed by a large number of "boneshaker" front drivers from about 1860. In the 1870s metal construction was mainly adopted, but there was a regular resurrection of the use of wood in various forms, including bamboo tubes, until the end of the century (Figure 10.1). Some bicycles were shown at the Stanley shows of this period with completely wooden wheels fitted with pneumatic tires, an example of which is still on show in the Science Museum, South Kensington, London in the form of an early Columbia bicycle. Various examples of wooden-framed bicycles dating to the 1890s are still ridden by proud owners in veteran-cycle rallies in addition to the more common "boneshaker" type machines.

Although wood was used regularly up to the 1930s for the making of rims (both tubular and wired-on types; and wooden mudguards and seat pillars were also not unkown) the wooden frame did not appear again until the 1940s, when it was thought that there was a case for saving metal for the war effort. However, wood became more scarce than steel. An American example is on show in the Washington Museum.[2] A cane-framed bicycle appeared in Trieste in 1945; Wilde thought it to be a sound proposition and stated that the machine was rigid enough for satisfactorily riding up hills.[3]

Plastic moldings: Since the recent advent of relatively large moldings in plastics (sometimes fiber-reinforced) there have been several attempts to market molded-plastic bicycle frames. Figures 10.2, 10.3, and 10.4 show that to varying degrees

Figure 10.1
Bamboo-framed bicycle.
From reference 8, p. 287.

Figure 10.2
United States figerglass-
frame bicycle, 1963.

Figure 10.3
British plastic-frame
bicycle.

Figure 10.4
Dutch plastic-frame
bicycle. Courtesy
Plastics and Rubber
Research Institute.

these frames are rather bulky—a feature not found to such a degree in the bamboo and other wooden-framed bicycles marketed over the years. To some degree the lack of popularity of these molded frames has arisen from their bulky appearance. As new polymers and polymer-fiber combinations are developed, plastic frames will become less bulky, lighter and, in particular, stiffer. There are certainly advantages for general everyday use in a frame made from a material which is completely resistant to corrosion and is inexpensive.

Plastic tubular structures: The bulky shape of the plastic frame can be avoided if the frame is constructed in conventional lines using tubes fitted into joints. Only recently have nonmetallic materials been made which in tube form could approach metals if weight and bulk were taken into account. Such materials are plastics reinforced with carbon fibers. These fibers are now made and sold at reasonable prices and have tensile strengths better than strong steels and high Young's modulus values. The fibers do not, however, exhibit one of the desirable properties of metals in that they do not stretch appreciably before breaking; also, the composite fiber structures have different properties "across the grain" than "with the grain." Also, the fibers have to be embedded in plastics which are very weak by comparison, giving a composite of varying properties, mostly less attractive by far than those of the fiber. Although the properties of carbon fibers are well known, the properties of usable forms, such as tubes, are not. However, according to one major manufacturer, a "Grafil" composite tube weighs less than a light-alloy tube of similar strength. The most unattractive feature of these tubes is that the joints have to be in the form of clamps. Adhesive joints are considered too weak and any drilling and riveting is liable to cause failure without warning. An example of a bicycle frame fitted with the main tubing in the form of carbon-fiber-reinforced plastic tubes is shown in Figure 10.5.

Figure 10.5
Carlton frame with
carbon-fiber-reinforced
plastic tubes and alumi-
num lugs. Courtesy Raleigh
Industries, Ltd.

**Frames in metals other
than steel**

In spite of the fact that very light steel frames of
about five pounds weight and less have been made
and used satisfactorily over a long period of cycling
history in various arduous circumstances, innovators
persist in advocating the use of other metals for
frame construction. It appears that weight saving
is the main object with a bonus in that the acceptable competitors of steel are corrosion resistant.
In general the costs are several-fold higher than for
steel frames of similar performance, both because
of the inherently higher cost of the metal and because of higher manufacturing costs.

Aluminum: The first innovations in nonferrous
metals for frames were introduced in the 1890s.

Aluminum was used both in the tubular form by Humber (with lugs which gripped the tube-ends) and by the manufacture of the Lu-mi-num bicycle made in France of cast alloy.

The Beeston Humber frame was reported by Wainwright as very satisfactory with a statement that the whole machine—with gear-case, lamp, and tools—weighed only 27 lbm.[4] There are no easily available records about the Lu-mi-num machine but a table of tests (Table 10.2) published in the journal *Engineering* of the period gives strength comparison with a current steel frame.[5] This shows that the balance for strength was in favor of the steel frame.

Since the introduction of aluminum frames, many other types have appeared on the market from Continental factories. Because aluminum brazing was formerly not practicable, various designs of lugs have been used to grip the tubes via corrugations or internal plugs. The most recent clamping type of lug can be seen on the make

Table 10.2. Tests on the "Lu-mi-num" alloy frame.

	Carried	Crippled at
Static Load on Crank Bracket		
Steel	2,925 lbm	4,172 lbm
Luminum	2,775 lbm	3,623 lbm
Static Load on Saddle		
Steel	4,275 lbm	5,438 lbm
Luminum	4,219 lbm	4,344 lbm
Static Load along line of Chainwheel		
Steel		2,600 lbm
Luminum		1,750 lbm
	Yielded at	Failed at
Load on One Pedal		
Steel	845 lbm	1,268 lbm
Luminum	300 lbm	1,250 lbm
Impact Test - Horizontal Blow on Front Fork		
Steel	3,544 lbm	4,463 lbm
Luminum	1,273 lbm	1,575 lbm

Source: Reference 5.

named Caminargent of the 1930s which used octagonal tubing. In addition various welded-joint frames have been marketed in spite of the bad effect which heating has on the properties of heat-treated metals.

Aluminum alloys, like most other nonferrous materials, do not in general have a "yield point" (a relatively large degree of stretch) before final failure, and bicyclists frequently report sudden failures of handlebars, handlebar stems, and seat posts which tend to discourage them from using aluminum alloy more generally. It is probable also that the manufacture of the special lugs as fitted to the Humber and Coventry Eagle (of 1930s date) were very expensive and the whole frame in general weighed but little less than a steel frame. (It will be noted that the Young's modulus values for aluminum alloys are one-third of that of steel, so thicker-than-expected tubing had to be used to give rigidity.)

Nickel: Nickel tubing followed the use of aluminum in the 1890s, no doubt in an attempt to produce a "rustless" frame. The firm manufacturing the frames, however, existed for but a short while during the bicycle-boom period when cost was of less importance. Nickel was and is an expensive metal compared with steel, but it is both strong and rigid and can be welded satisfactorily. It is seldom used in its pure form but is a major component, with chromium, of stainless and high-strength steels.

Titanium: The history of the use of aluminum for bicycle frames is repeating itself in the case of the recent use of titanium. Within a decade of the commercial production of a once very costly metal it is being thought seriously of as a usable metal for bicycle frames.

Titanium in various alloy compositions is now available for corrosion-resistant heavy engineering equipment and for high-speed aircraft and compressors. Satisfactory welding methods using inertgas shielding to avoid weld deterioration have been developed, and suitable tubing of about the tensile strength of steel can be so joined.

Titanium has a specific gravity about half that of steel and is, for bicycle usage, corrosion proof. As a consequence it was possible for the firm of Phillips to produce a fairly conventionally shaped bicycle frame weighing 2¾ lb [1.25 kg] and to put it on show at the London Cycle Show in 1956. No models were offered for sale—the price would have been high. The Speedwell Gear Case Co. Ltd of Birmingham is currently (1973) producing 10,000 frames with titanium tubing and selling them for about £130 or the equivalent. The mass of the frame and fork was advertised as 3¾ lb [1.7 kg] which is heavier than the Phillips' frame.

To date there is no published information about riders' opinions of the riding characteristics of titanium frames. The ultimate tensile strength of the tubing used is probably similar to that of the standard steels used for frames, but the Young's modulus values for titanium are one-half of that of the steels. Hence, unless the structure is designed differently or incorporates extra thicknesses where required, it should be less rigid.

Magnesium and beryllium alloys: The only other metals likely to be considered for bicycle-frame construction are magnesium and beryllium alloys. The former are well developed and have an attractively low specific gravity of about 1.7, which to some extent compensates for the relatively low tensile strength and very low Young's modulus values which are one-fifth of that of steels. An alloy termed Elecktron was used fairly satisfactorily for making bicycle rims in the 1930s, but there have been no further applications in cycle manufacture. Beryllium is a lightweight metal of specific gravity 1.85 and is not in the same advanced state of development. Reports to date emphasize the possible saving of weight as compared with similar aluminum-alloy structures but stress also the low ductility, high cost, and poor machining qualities.

Conclusions and speculations

Although much experience has been obtained with the manufacture of bicycle frames and accessories in steel and aluminum alloy along with the production of low-stress parts, such as mudguards (fenders), in relatively well-known nonmetallic materials, there seems an urge to try out new and expensive substances. The aims appear to be directed on the one hand toward producing a lighter and more corrosion-resistant product and on the other to making a unit frame instead of an assembly of parts in an inexpensive and again corrosion-resistant nonmetallic material. There might be advantages in using other unit-construction methods for metals which avoid machining, such as lost-wax precision casting. This process is ideal for mass-production purposes.

We can expect improvements in frame design and manufacture to give greater torsional stiffness.[6, 7] Such improvements would enable an acceptable one-piece carbon-fiber-reinforced-plastic standard-type frame. This could avoid the use of bulky and weak joints and take full advantage of the fiber strength.

Design considerations, present and future: Optimized designs for a lightweight bicycle wheel and frame were evolved by the 1870s. The tension-wire-spoked suspension wheel and hollow-metal-membered brazed-joint frame had by then ousted all others. These designs were pioneered by the cycle industry and not copied from some other branch of current engineering practice. This pioneering led to the establishment of specialized industries, such as steel-tube manufacture, and in addition an accelerated progress in the ball-bearing-manufacturing industry, and so, in a significant way, helped to launch the aviation age.

Current engineering science and practical considerations established the closed-section frame member mostly of round or near-round section assembled with rigid joints generally incorporating lugs. There was a rapid rejection of the practice common to other structural-engineering practice of using channel-section members bolted at the joints. The optimum frame shape also came early in the progress of the bicycle industry in the form of the Humber pattern, now called the diamond frame. Before this standardization came about in the 1890s there had been a multitude of frame patterns, mostly constructed of much more robust and heavier tubing than that desired in the 1890s when only reasonably lightweight machines were acceptable. Some of these early designs of frame appear nowadays incorporated in childrens' machines or special machines for adult use, such as the modern small wheelers.

Evidence that the diamond shape of frame is for most purposes based on sound distribution of constructional metal is given by the fact that lightweight track racing bicycles can be built that weigh but 6¾ to about 10 pounds [3-4.5 kg] (Figures 10.6, 10.7, and Table 10.3). The steel-tubed frames of these machines must be very lightweight indeed, showing that there is a good approach to a minimum of metal and hence an optimum placing of the members. It is interesting to compare the above weights with the weight of 8½ pounds [3.85 kg] of a pair of pneumatic-tired roller skates shown in Figure 10.8. These skates represent a high degree of "weight paring" for a wheeled man-propelled machine, and yet complete bicycles have been made lighter.

From the facts given above it appears likely that we already have a very nearly optimum shape for bicycles which assures good distribution of material, minimum weight, and (less obviously apparent) low wind resistance. If, therefore, the goal of future designers is any further decrease in weight and/or greater rigidity or general durability, they must look for different materials than steel and aluminum alloys. Even if cost is disregarded the designer is placed in a quandary when attempting to use new materials, because he has but little latitude to alter the present designs, the refuge allowed to most designers of engineering structures when contemplating a change in materials of con-

Figure 10.6
Lightweight bicycle of
1895. Perhaps the lightest
standard, adult-size bicycle
ever built was this Tribune
of 1895, exhibited at the
National Bicycle Exhibition
at Madison Square Garden,
New York, of that year.
This featherweight weighed
eight pounds, fourteen
ounces, ready to ride.
Reproduced with permis-
sion from *Riding high: The
story of the bicycle* (New
York: E. P. Dutton &
Company, 1956).

Figure 10.7
The Raleigh Professional
Track Cycle, 1974.
Courtesy Raleigh Industries
of America, Inc.

Figure 10.8
Pneumatic-tired roller
skate.

Table 10.3. Lightweight steel-frame bicycles.

Date	Weight, lbf	Type	Make	Material of construction
1888[a]	$15\frac{1}{2}$	Cross-frame safety, solid rubber tires	Demon	Steel
1888[a]	19	Diamond-frame safety, solid rubber tires	Referee	Steel
1888[a]	11	High ordinary	James	Steel
1895[b]	$8\frac{7}{8}$	Diamond-frame safety, Pneumatic tires	Tribune	Steel
1948[c]	$8\frac{5}{8}$	Modern track bicycle	Legnano	Steel & Alloy
1949[d]	$6\frac{3}{4}$	Modern track bicycle	Rochet	Steel & Alloy

Sources:
[a]*Bicycling News,* 8 February 1888.
[b]*Riding high: The story of the bicycle* (New York: E. P. Dutton & Company, 1956), p. 125.
[c]*Cycling,* 7 January 1948, p. 10.
[d]*Cycling,* 3 November 1949, p. 514.

struction with less desirable properties than the original. A low Young's modulus value for a material could be compensated for by the use of deeper sections of members. To some degree all proposed new materials for bicycle construction have low Young's modulus values with the exception of carbon-fiber-reinforced plastic tubing. This latter, however, poses most difficult jointing problems. The excessively deep members necessary for a sound plastic frame are illustrated in Figure 10.4. From the point of view of wind resistance alone these cannot be said to be optimum. Frames in titanium can be of a shape very similar to the standard steel pattern because the tubing thickness could be increased without disturbing the outward shape of the frame. Here again a precise balance must be reached between metal thickness and maybe tube-outside-diameter increase to make a member as rigid as a steel member. Otherwise the member in the new material could be heavier than the steel counterpart.

Many examples are given by Sharp[8] concerning the calculation of stresses in cycle frames for the cases of the more simple static loadings. It appears that the use of standard modern lightweight strong steel tubing of near 22 gauge provides a safety factor above the yielding point of about three for distortion of the bracket through full-weight pedaling. The safety margin for simple vertical loading on the saddle pillar is very large, being some ten times the threefold previously mentioned. (See Table 10.2 for actual tests on frames.)

References
Chapter 10

1. I. Cohen, "Polytetrafluoroethylene," *Cycling,* 24 March 1955, p. 301.

2. Hempstone Oliver Smith, "Catalog of the cycle collection of the division of engineering," U.S. National Museum, Bulletin 204, Washington, D.C.: U.S. Government Printing Office, 1953.

3. J. Wilde, "A cane bicycle from Trieste," *Cycling,* 22 December 1945, p. 420.

4. G. S. Wainwright, "Aluminum cycles," *CTC Gazette,* July 1896, p. 311.

5. A. C. Davison, "The Lu-mi-num frame," *Cycling,* 19 February 1941, p. 157.

6. "Technology transfer at B. S. A," *Engineering* (London), vol. 210, 22 January 1971, p. 736.

7. "Introducing science to a craft," *Engineering* (London), vol. 212, April 1972, p. 374.

8. A. Sharp, *Bicycles and tricycles* (London: Longmans, Green & Company, 1896).

Additional recommended reading

Bartleet, H. W. "An early aluminum bicycle," *Cycling,* 11 February 1942, p. 113.

Couzens, E. G. and Yarsley, V. E. *Plastics in the modern world* (Harmonsworth, Middlesex, England: Penguin Books, 1968).

Design engineering plastics handbook (West Wickham, Kent: Summit House Glebe Way).

Gordon, J. E. *The new science of strong materials* (Harmonsworth, Middlesex, England: Penguin Books, 1968).

Whitt, F. R. "Alternatives to metal," *Cycle Touring,* June-July 1971, p. 138.

Whitt, F. R. "Bicycles of the future," *Cycle-Touring,* August-September 1967, pp. 155-156.

Part III   Other human-powered machines

11

Unusual pedaled machines

As has been stated before, it is probable that widespread development of better roads made the use of bicycles much more practical. The propulsive power needed was then brought below that for walking or running at comparable rates and the encumbrance of a machine became justifiable. Although walking on soft ground requires a twofold increase in effort compared with that needed on concrete, some fiftyfold increase in resistance is experienced by a wheeled machine. So on soils, the advantages of a wheeled machine to a walker are diminished.

Most of the roads covering the world are made of bonded earth with relatively poor surfaces, and as a consequence bicycle usage, in general, is under less-than-optimum conditions. The bicycles used on these roads are somewhat different from what is now the familiar pattern on good U.S. and British roads. Throughout the world, particularly where roads are poor, the 28 in. [about 700 mm] wheel with a tire of about 1½ in. [39 mm] cross-section is commonly used. Big wheels with large tires have also provided a partial solution to an as yet unsolved optimum design for agricultural and military vehicles, which have to travel on poor surfaces.

In addition to attempting to solve the problems associated with the use of vehicles on poor roads, inventors have tried to devise man-powered vehicles for other environments.

Machines for riding on water, on railways, or in the air have been the targets of inventors ever since the practical bicycle appeared in the late nineteenth century and demonstrated its speed on good roads.[1-4] It is probable that the bicycle's high efficiency under good conditions was taken, mistakenly, by inventors to imply that similar performances could be expected from its use in very different conditions.

Water cycles

Through the building of hard, smooth-surfaced roads, man has been able to use to his great advantage wheeled machines in order to progress with the minimum of effort. It is not possible, however, to duplicate this achievement with water surfaces and produce "smoother water." The resistance to movement offered by a relatively dense and viscous medium such as water is great compared with that offered by air. As a consequence, both submerged and floating objects, such as swimmers and row boats, can travel at only a quarter the speeds (at similar effort) of their land counterparts, runners and bicyclists.

Inventors have persevered over the years, however, and many watercycles have appeared. Modern versions are seen at seaside resorts. The form is often that of side-by-side two-seater pedaling machines, an arrangement long abandoned for serious tandem bicycle-type construction, although it was popular in the 1870s. An illustration of an early triplet water cycle is given in Figure 11.1.[5]

According to the *Dictionary of Applied Physics,*[6] screw propellers, paddle wheels, and oars can all be designed and used to give an applied power efficiency up to about 70 percent.

However, the kinetic energy imparted to oars

Figure 11.1
Triplet water cycle.
Courtesy of Currys, Ltd.

in the forward and return strokes is lost during rowing, and a large proportion of the thrust is at an angle to the direction of motion, both of which features constitute inefficiencies. Screw propellers have been able to exceed paddle wheels in efficiency by a considerable margin. Therefore the above figure of 70 percent for all three devices must be considered to be a rough approximation, because an optimized screw propeller can perform at a much higher figure.

The power absorbed by water friction on the hull of a streamline-shaped boat can be represented approximately by the equation

Power (hp) = 0.000024 $\times$ wetted surface (sq ft)

$\times$ [speed (knots)]$^{2.86}$,

or

Power (watts) = 1.287 $\times$ wetted surface (sq m)

$\times$ [speed (m/sec)]$^{2.86}$.

Some additions of typically 10 to 20 percent have to be made for imperfectly streamlined hull design and for wind resistance.

The wetted surface for boats and water cycles designed to carry the same weight can be similar. Hence it can be concluded that for a given power input by the oarsmen or pedalers, boats and water cycles propelled by screw propellers should travel at a slightly higher speed than those driven by oars or paddles, even though the differences will be small.

Some evidence of the validity of the above conclusion is provided by an account[7] of the performances of water cycles in their heyday of the 1890s. A triplet water cycle ridden by the ex-racing bicyclist F. Cooper and two others covered 101 miles [162.5 km] on the Thames from Oxford to Putney in 19 hours 27 minutes and 50 seconds. A triple-sculls boat rowed by good university oarsmen covered the same course in 22 hours and 28 seconds. The water cycle was the faster vehicle by about 18 percent.

Other facts about water cycles in this period are interesting. The English Channel was crossed, Dover to Calais, by a tandem water cycle in 7¼ hours. A sextuplet water cycle ridden by girls on the Seine is credited with reaching a speed of 15 mile/h [6.71 m/sec]. "Hydrocycles" were manufactured by L. U. Moulton of Michigan, and said to be capable of speeds of 10 mile/h [4.47 m/sec]. All these performances compare favorably with oar-propelled boats rowed by the best oarsmen.

In order to permit riding in water, "amphibious" machines have been constructed and ridden. These had floats which were so arranged that when the machine was ridden on land they did not obstruct its movement.[8]

Ice and snow machines

In addition to water cycles, attempts have been made to develop and popularize bicycle-type machines for running on ice or on snow.[9] Some types consist of a bicycle with a ski replacing the front wheel. Others dispense with wheels and retain only the frame, with two skis attached, one on either side. Unlike the case of water cycles, there is no published evidence concerning the speediness of these machines compared with skating or skiing.

Railway cycles

The resistance to motion offered by a steel wheel running on a steel rail is very low indeed and less than that of the best of pneumatic-tired wheels running under optimum conditions of road use. As a consequence, cycles developed for running on rails have been proved practical in the sense that they were not difficult to propel. In fact, high speeds are credited to this type of machine. An illustration of one type is given in Figure 11.2.[10]

A drawback to railway cycles is the general unavailability of unused lines; the Victorians took quite seriously the idea of laying special cycle tracks alongside the regular rail tracks in some areas.

Air cycles

The dream of man-powered flight has inspired inventors in the past. It will probably continue to fire the imagination of men for some time yet.

Figure 11.2
Early railway cycle.
Courtesy of Currys, Ltd.

Since at least 1400 B.C. attempts have been made to fly, unaided, by all types of constructors, both serious and maniacal; the challenge has proved irresistible.

The design of high-powered airplanes progressed so rapidly after 1904 that the science of low-powered flight was not, as might be reasonably expected, fully explored. As a consequence teams and individuals are, even now, engaged in unraveling the scientific problems associated with flight at low speeds and close to the ground. The whole process has been greatly accelerated by the promise of a prize ($120,000) for the first flyer(s) to complete a figure of eight over a distance of one mile.

The terms of the Kremer prize preclude the use of bouyancy such as that given by a balloon or airship. So, like other modern flying machines, the man-powered machine must use power in keeping itself and occupant(s) aloft. This is the great difference between progress on a supporting solid surface and flight through the air where an upthrust, as well as the force to move forwards, must be developed by the propulsion unit.

The information published so far has appeared, mainly, as short articles in the daily press.[11-25] As might be expected, with a competition still on for a large prize, constructional details are often kept secret.

In general, the latest types of man-powered airplane differ from those tried out in the early 1900s and referred to by an observer of that period, G. H. Stancer.[26] (Figure 11.3). He wrote as an observer of the trials in France of 199 entries that only some short "jumps" were attained. Modern designs include a machine with an inflatable wing and at least a couple of two-man-power machines. A present-day two-man machine being developed by students at the Massachusetts Institute of Technology is shown in Figure 11.4. Also the size of the machines is much greater than in the pre-World War I types, indicating a much lower lift pressure from the wing surfaces. Even today there is a difference of opinion about pusher and tractor types of propellers. Some helicopter types are still being constructed although, as yet, none have flown. A flapping-wing machine also exists.

A machine developed in Germany in the 1930s "jumped" 790 yards [720 m], from an assisted take-off,[10] thus achieving much more than did those of the early 1900s performing at the Parc de Princes trial. The latest efforts in England are also better performances.[27]

A machine being built at Southhampton University made its first successful "jump" of 50 feet during its early stages of development.[28] A week later, in 1961, Puffin Mark I, from the Hatfield Man-Powered Aircraft Club and flown by John Wimpenny, flew 50 yards at a height of 5 feet. An improved model in 1962 flew the greatest distance to date, about half a mile, in 2 minutes. Neither machine could attempt the turn required by the Kremer Prize conditions. These machines have large wing spans of about 90 feet, which is similar to that of medium-sized airliners. The propellers of the successful machines are likewise large.

Among the other promising designs from the 20 or so types being developed is that of a group at the University of Belfast, headed by Professor

Figure 11.3
Early air cycle.
Courtesy of Currys, Ltd.

Figure 11.4
The MIT man-powered
aircraft. MIT photo by
Calvin Campbell.

T. Nonweiler.[29] This is, like several other types, a two-seater. So much concentration on controlling is required to execute the turns and other maneuvers specified by the Kremer prize conditions that it is considered better by this group that a "partial" passenger be carried for this purpose, preferably one with piloting experience. In addition, the weights of the machines, although light (75 lbm to about 160 lbm [34 to 73 kg]), are a sizable proportion of that of a man and therefore with two men, a greater pro rata power per unit weight is achieved.

It appears that sky cycling is not likely to be a cheap sport.[30] The cost of developing a machine, with some expectations of it being able to fly, is likely to be about $250 per pound weight of machine. This cost is more than that of standard airliners. Cost figures given for some single individual efforts, as distinct from products of teams, are, however, much lower. It is likely that any satisfactory machine will be comparatively large, so its storage and use will be limited to that provided by an airfield. As a consequence, extra costs for use will be involved, in addition to that of the machine itself.

Although the development of air cycles is probably one of the least utilitarian types of endeavors in the history of bicycle adaptation, it is technically one of the most interesting. The latest upsurge of activity forms an as yet unclosed chapter in the bicycle's long history.

A pedal-driven lawn mower

The rationale behind the design of the grass mower shown in Figure 11.5 was that the leg muscles would be used more efficiently in pedaling than in pushing a regular lawn mower, and the back and arm muscles would be relieved; that continuous mowing would be more efficient than the frequently used to-and-fro motion of push mowing; that a multiratio gear would enable individuals to choose whatever power-output rate suited them and would enable moderate slopes to be more easily handled; that shortages of gasoline and antinoise

Figure 11.5
Diagram of the Shakespear
pedaled mower. From
reference 31.

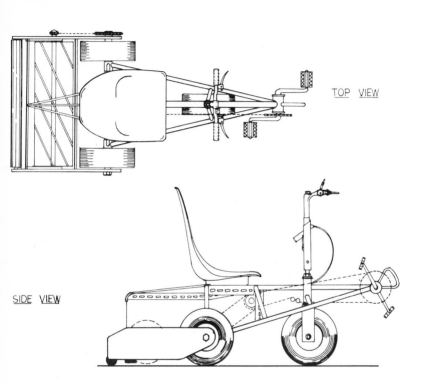

TOP VIEW

SIDE VIEW

restrictions might limit the use of power mowers; and that riding a pedaled mower might be fun as well as good exercise.

The original model shown in Figure 11.6 was designed and constructed by Michael Shakespear of the Massachusetts Institute of Technology for his mechanical-engineering thesis.[31] A three-speed Sturmey Archer hub gear, a brake, and differential were incorporated into the transmission. The reel-type cutter is driven directly from the input to the differential drive to the rear wheels. Pulling the left-hand handlebar lever releases a catch and enables the cutter assembly to be raised by pulling the handlebars back to a rear position and so permits easy maneuvering. The prototype, constructed largely of scrap materials and components, was very heavy but still gave easy cutting. A lightweight model might show real advantages.

Energy-storage bicycles The concept of storing braking or downhill energy, or even energy pedaled into the machine while the rider waits at a traffic light, and then drawing cn the stored energy for burst of power up hills or for acceleration, is one which has intrigued inventors for many years. In every bicyclist there is a desire, however suppressed, to leave the sports cars standing in a cloud of rubber smoke. Sad to state, the chances are small.

Table 11.1 shows some maximum energy-storage capabilities of various systems. [32] Flywheels are so much better than rubber bands or springs that they would be the preferred contenders, and they have many enthusiasts.[33] (City buses driven by flywheels were manufactured by Sulzer in Switzerland; the flywheels were speeded up by electric motors at stops). Compared with the energy-storage capability of gasoline, however, a flywheel is almost 100 times heavier. And it needs an infinitely variable transmission if its kinetic energy is to be transferred efficiently to the driving wheel. The "windage" losses constantly degrade the stored energy. All these factors mean high weights and

Figure 11.6
Michael Shakespear on his
pedaled lawnmower.

Table 11.1. Maximum energy-storage capability.

A. Maximum energy-storage capability of various materials

	Electrochemical[a] conversion, watt h/lb	Heat-engine[b] conversion, watt h/lb	Mechanical conversion, watt h/lb
Hydrogen[c]	14,900	3,040	
Gasoline[c]	5,850	1,130	
Methanol[c]	2,760	505	
Ammonia[c]	2,520	503	
Hydrogen-oxygen (liquid)	1,660	338	
Lithium-chlorine (700° C)	980		
Magnesium-oxygen[d]	1,800		
Sodium-oxygen[d]	775		
Zinc-oxygen[d]	500		
Sodium-sulfur (300° C)	385		
Lithium-copper-fluoride	746		
Zinc-silver dioxide (silver-zinc-battery)	208		
Lead-lead dioxide (lead-acid battery)	85		
Cooling lithium hydride		64	
Flywheel			14
Compressed gas and container			10
Rubber bands			1
Springs			0.06
Capacitors			0.006

B. Energy density in kilowatt h/cu ft.

System	kW-h/ft^3 Low	High
Electrostatic		0.0045
Magnetic	0.0007	0.06
Gravitational	0.006	0.15
Mechanical	0.0007	0.6
Phase change	0.007	75.
Primary battery	0.15	7.5
Secondary battery	0.45	1.5
Fuel cell	0.75	75.
Fuel		300.

Source: Reference 32, p. 54
[a]Based on Gibbs free energy
[b]Assumes 20 percent thermal efficiency
[c]Reaction with oxygen from atmosphere
[d]Including weight of oxygen

high losses, neither welcome for bicycle components.

Batteries are better as far as weight for the energy storage alone is concerned. But then a motor and control system and transmission are required. At least a half-horsepower capability would be desirable, and a minimum weight for a special motor and transmission might be ten pounds. The battery and housing would be another ten pounds. (Extremely expensive aerospace-type components would be required to keep weights down to these levels.) A lightweight bicycle would about double its weight, and the rider could well feel that he might as well go a step further and have a regular— or even a battery-powered—motorcycle.

These conclusions have been given some weight by a study performed by students at Dartmouth College, Hanover, New Hampshire.[34]

They adopted a practical outlook and devised a specification which included a price of $50, a weight of 30 lbm [13.6 kg] and a power output sufficient to propel the rider and machine up a hill of length 2120 ft [645 m] and height 90 ft [27.4 m].

Four systems were studied: a spring, a flywheel, electrical storage, and hydraulic storage. It was decided that there was no spring system which could be described as practical. The hydraulic system would cost $1500 and would have to work at extremely high pressures, resulting in a large weight.

The mechanical flywheel system they calculated would be suitable if it incorporated two 35 lbm [15.9 kg] flywheels revolving at 4,800 rpm, characteristics right outside the specifications.

The electrical system of motor/generator and electricity accumulator would cost $74 and have an overall efficiency of 34 percent and weigh 40 lbm [18.1 kg]. This is much nearer to the specification. However, the low efficiency and high weight and cost make the concept very unattractive.

For further discussions of energy storage and of earlier attempts, see references 35 and 36 and Figures 11.7 and 11.8.

Figure 11.7
Racing bicycle with
flywheel.

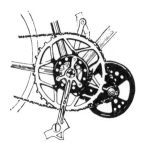

Figure 11.8
Thompson flywheel
mechanism.

**References
Chapter 11**

1. *The Rambler* (London: Temple House E. C. I., 1897).

2. *Strange but true,* Curry's Ltd., nos. 12, 26, and 48, 1940.

3. A. J. Palmer, *Riding high: The story of the bicycle* (New York: E. P. Dutton & Company, 1956).

4. John Hadfield, *Saturday book* (Boston: Little-Brown, 1965).

5. See also references 1, 2, and 3 above.

6. Sir Richard Glazebrook, editor, *A dictionary of applied physics* (London: Macmillan & Company, 1922).

7. See reference 1 above.

8. See references 2 and 3 above.

9. See reference 3 above.

10. See also references 2, 3, and 4 above.

11. Various articles in *Cycling,* 1909-1912.

12. "Bicycle-powered flight," *Cycling,* 3 June 1959, p. 382.

13. "Flying bikes on the way," *Reveille,* 30 April 1959.

14. Helmut Haessler, "Man-powered flight in 1935-37 and today," *Canadian Aeronautical Journal,* vol. 7, March 1961.

15. John Davy, "Taking to the air on a tandem," *Observer* (London), 1 November 1959.

16. "The sky bike," *Cycling,* 4 November 1959, p. 21.

17. "Pioneering the air bicycle," *Observer* (London), 18 June 1961, p. 12.

18. A. Macpherson, "Meet the sky bike," *Daily Mail* (London), 7 November 1961.

19. "The puff-puff puffin," *Daily Mail* (London), 22 November 1961.

20. "The pedal plane," *Daily Herald* (London), 5 May 1962.

21. Joan Green, "Flycycling this year," *Cycling,* 23 May 1962, p. 14.

22. John Davy, "Pedalling across the sky for £5,000." *Observer* (London), 28 June 1964.

23. M. Kienan, "Man powered plans are ready," *Sunday Times* (London), 25 April 1965.

24. "Daring young men and their flying machines," *Daily Sketch* (London), 19 July 1965.

25. M. Mayntham, "The back-parlour bird man works by gaslight," *Sunday Times* (London), 24 July 1966.

26. G. H. Stancer, "Revival of the cycloplane," *Cycling,* 25 November 1959, p. 9.

27. See reference 24 above.

28. See reference 22 above.

29. T. R. F. Nonweiler, "Man-powered aircraft: a design study," *Journal of the Royal Aeronautical Society,* vol. 62, 1951, pp. 723-734.

30. See references 16 and 22 above.

31. Michael Shakespear, "A pedal-powered riding lawn mower," B.S. thesis, mechanical engineering department, Massachusetts Institute of Technology, Cambridge, Massachusetts, June 1973.

32. John F. Kincaid et al., "The automobile and air pollution: a program for progress, part II," report PB 176 885, U.S. Department of Commerce, Washington, D.C., December 1967.

33. R. F. Post and S. F. Post, "Flywheels," *Scientific American,* vol. 229, December 1973, pp. 17-23.

34. "Report on the energy-storage bicycle," Dartmouth College, Hanover, New Hampshire, 1962.

35. See reference 3 above.

36. F. R. Whitt, "Freewheeling uphill—is it possible?" *Cycling,* 30 January 1965, p. 13.

Additional recommended reading

History of aviation, part 6 (London: New English Library Ltd., 1969).

"Pedal-power flight beaten by wind," *Daily Telegraph* (London), 20 March 1972.

Rouse, H. *Elementary mechanics of fluids* (London: Chapman & Hall, 1946), p. 286.

Shepherd, E. Colston. "What happened to man-powered flight," *New Scientist,* 27 November 1969.

Sherwin, Keith. *Man-powered flight* (Hemel Hempstead: Model and Allied Publications Ltd., 1971).

Worth, Paul. "Man-powered planes get a new lift," *Popular Science,* 1972, pp. 67-68.

Man-powered vehicles in the future

Bicycling as a means of transport rose rapidly to an almost incredible level of popularity in the 1890s, as has been stated earlier. Many roads were either created or paved as a direct result of the bicycle "craze." There was an outpouring of creative talent, and the design of man-powered vehicles went through almost every possible variation before the combination of the pneumatic tire and the "safety-bicycle" configuration showed such clear superiority over other contenders that it has reigned unchallenged since.

Indeed, there have been very few changes to the design of the standard bicycle since 1890. The reason for this is not entirely that the safety bicycle represented the ultimate in man-powered vehicles. It is, rather, that the appearance on the transportation scene of the internal-combustion-engine-powered automobile siphoned off all the adventurous mechanical engineers and backyard mechanics into that field. Almost carbon-copy bicycles, million after million, have been made since that time, with changes no greater than minor variations in wheel diameter, tire diameter, frame angles, and gear ratios.

Where the automobile is out of the reach of the pocket of most people, the bicycle still reigns supreme—over much of Africa and Asia and some of Europe. The Viet Cong were supplied by trains of bicycles. In Nigeria a bicycle was, and probably still is, a highly prized possession, often taking precedence over a wife, whose purchase price was often comparable.

The present picture in the United States

America is a nation on wheels: by this trite phrase one means that there are nearly one-hundred million motor vehicles on the roads. There are also about ninety-million bicycles (1973) in the United States. While one reason for this high figure is the affluence which enables a person to buy a bicycle even if he does not intend to use it every day, it is still true that bicycling is the fastest-growing sport for competition and recreation in this country. Many cities and states have designated "bikeways" following the example of the initial bikeway in Homestead, Florida. When Mayor Lindsay's Commissioner of Parks closed Central Park in New York to all but bicycles on Sundays, the response was so large that it had to be concluded that a much larger proportion of the population than is generally assumed would enjoy daily the gentle exercise of bicycling if it were not for the constant danger and unpleasantness of competing for space on the roads with high-powered cars.

The bicycle, and possible future vehicles

The bicycle is, in good weather and on smooth roads, a truly amazingly convenient means of transport. It gives door-to-door instantly available service at an average speed in urban areas usually better than that of any competitor, at least for distances up to five miles. It is extraordinarily light (payload up to ten times the unladen weight) and narrow, so that it can travel and be stored in places inaccessible to motor vehicles. A bicycle can pay for itself in saved fares in much less than a year. And, of course, it is an almost perfect way of getting exercise and keeping healthy.

All these attributes of this wonderful vehicle have been with us since before the turn of the century. So have nearly all of its shortcomings, some of which are listed here.

1. The braking ability of bicycles is very poor, especially in wet weather.

2. A bicycle rider, unless he wears cumbersome clothing, is unprotected from rain, snow, hail, road dirt, or from injury in minor accidents.

3. It is difficult to carry packages, briefcases, shopping bags, etc. conveniently or safely.

4. The aerodynamic drag in a headwind is very high.

5. The riding position and the pedal-crank

power input are not ergonomically optimum.[1]

6. The reliability of bicycles is very poor (especially with regard to brake and gear cables and wheel spokes) and in regard to maintenance its present design is attuned to the low-labor-cost conditions of an earlier age.

7. Whereas family cars retail at about 75¢ per pound, regular bicycles cost about $2.00 per pound (and lightweight models may easily cost $20.00 per pound), although they contain much less sophisticated engineering than do automobiles.

The correction of these drawbacks would provide little problem to NASA or General Motors. Their continued existence is the consequence of a vicious circle having developed. This vicious circle is similar to that which has caused the running down of public transportation: too many cars led to such unpleasant conditions for bicycling that demand slackened; manufacturers cut out all "nonessential" expenditures; and nineteenth-century bicycles made poor competition with highly developed modern automobiles.

The situation may be changing. The unhappy state of our cities, the at-last-recognized harmful effects of automobile congestion in urban areas, the growing shortages of energy and raw materials, the concern over the damage to our environment— all of these are helping to recruit not only new bicyclists but also scientists and engineers anxious to solve problems.

Man-powered land transport competition

Some of the new developments being continually reported in the man-powered-transportation field may have been partly inspired by an international competition organized in 1967-1968.[2,3] The aim of the competition was to encourage improvements in any aspect of man-powered land transportation.

The search for an improved vehicle may well start from an appreciation of the good and the bad qualities of the present bicycle as listed above. The shortcomings mentioned are, of course, generalizations purposely made more negative than is always warranted, although some competitors

Figure 12.1
Enclosed bicycle with
outriggers.

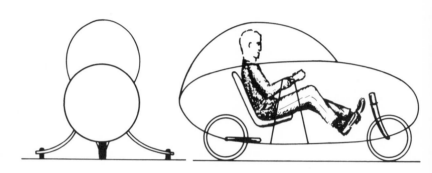

were much harsher than this in their criticisms. Let
us look at some of the possibilities of overcoming
these objections and at the suggestions made by
some of the competitors.

There were many proposals incorporating
bodies to give weather and minor-accident protec-
tion and luggage space, combined in some cases
with a reduction in air drag in a head wind. Some
entrants recognized the penalties in increased
weight, of side force in a cross wind, and of usually
more difficult access to enclosed vehicles. The
bodies were virtually all added to a chassis or
spine rather than being designed to supply struc-
tural strength. No one experimented with a "crus-
tacean" rather than a "vertebrate" construction;
in this the competitors were probably wisely conservative.

Whether the advantages given by a body can
justify its drawbacks will be known only through
public acceptance. Most riders would not like to
sacrifice the bicycle's narrow width and its ease of
maneuvering and parking, but many would be well
prepared to accept a weight penalty of 15 lbm

[6.8 kg] in a commuting vehicle if the body would keep the rider (and briefcase) clean and dry, warm in winter, and as cool as possible in summer. There were several "bodied" entries that met most of these criteria, though few were greatly concerned with the weight reduction which would seem desirable.

Many competitors felt that it was logical to combine a body with a tricycle or four-wheeled configuration. Obviously there is an immediate addition of weight and of width for stability if only because the wheels and suspension must now handle high side loads that are absent from bicycles. If we set out to attract a housewife, perhaps with a baby, to go shopping under her own power, we might find that a three-wheeler or four-wheeler (which has one more wheel but one less track than the usual tricycle) would have a great appeal. The additional vehicle weight, at least, matters less when one is carrying cargo.

A configuration which might have advantages is that of a two-wheeled single-track vehicle with a "feet-up" body and outriggers which could be dropped when one stopped (Figure 12.1). And for a three-wheeler the arrangement of a motorcycle and sidecar gives two tracks instead of three, and might have other advantages.

The body shape, rider attitude and wheel arrangement are intimately connected with the power transmission, and in this area competitors spent much creative effort. There was much preoccupation with constant-velocity foot motion in a straight line or through an arc.

Some entries proposed hydrostatic transmission which would at least give efficient braking on the driven wheel and possibly an infinitely variable gear ratio. The weight penalty, however, would be severe.

There was little evidence of much emphasis being given by competitors to the severe problem of braking in general. About an equal number of competitors ascribed the poor performance of the rim brake in wet weather to high brake pressure

Figure 12.2
Lydiard "Bicar"—Mark III.
A half-reclining position of
the rider is adopted in the
Bicar. Swinging cranks
actuate through pull rods
and the rider puts his legs
through flaps in the body
to rest on the ground.
Towing tests indicated
average touring speed may
be increased by 6 mile/h.
From reference 3.

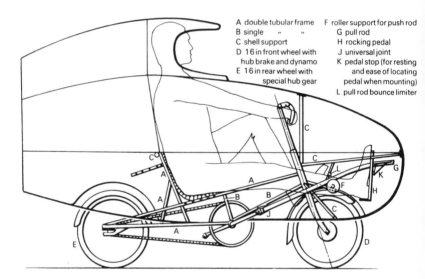

A double tubular frame
B single   "   "
C shell support
D 16 in front wheel with
   hub brake and dynamo
E 16 in rear wheel with
   special hub gear

F roller support for push rod
G pull rod
H rocking pedal
J universal joint
K pedal stop (for resting
   and ease of locating
   pedal when mounting)
L pull rod bounce limiter

as to low, but no one carried out the simple tests
needed to determine the point. No one suggested
any form of servo assistance from the wheel motion
to reduce the cable tension required. There were
several designs on paper of drum or disk brakes,
but nothing to suggest that they would be any
better than present brakes and much to indicate
a substantially higher cost. The judges were disap-
pointed at the lack of brake developments because
they would have given the first prize to anyone
who had made or modified a brake to give improved wet-weather operation and higher cable
reliability without adding greatly to the weight
or cost.

Rim brakes virtually necessitate metal wheels,
but there were several proposals for unspoked
wheels which were occasionally to be of plastic,
plywood, or dished aluminum. Many competitors
did not appreciate what a great advance the
"tension" spoked wheel was when it was introduced and that wheels would almost inevitably
require a greater weight if components in bending
and compression were substituted. Glass fibers in
tension held in resin might be a good substitute
for spokes and might give a lighter and more robust
wheel, suited to mass production. A metal hub and
rim for tire retention and for braking would probably still be needed.

The Bicar

The first prize went to W. G. Lydiard who, besides
carrying out careful design and analytical work in
the areas of stiffness, stability, aerodynamics, transmission, and so forth, made three experimental
machines of different configurations. His first
model was a three wheeler, the other two had two
wheels of 16-in. [40.6 mm] diameter (Figure 12.2).
Lydiard calls his Mark 3 machine (which he does
not claim to be near a final solution) the Bicar, a
name which correctly implies that the rider is
housed in a body and pedals in a half-reclining
position.

A problem identified by Mr. Lydiard with
two-wheel reclining-rider bicycles is that either the

wheelbase and overall length become excessive,
or the front legs must pedal over the front wheel.
He found that a conventional chainwheel and
cranks in this position gave a marked "feet-up"
attitude, and he eventually adopted pull rods swinging through arcs operating cranks in a more-or-less
conventional position. He found that these pull
rods interfered somewhat with his ability to put
his legs on the ground through flaps in the body,
and for a later machine he is proposing pull rods
operating possibly a variable-ratio over-running
gear in the rear wheel, together with sprung wheels
(Figure 12.3).

The Bicar's body is of 1 mm ABS plastic; Mr.
Lydiard intends to try ½ mm ABS to reduce body
weight and also, possibly, ¼ in. (63 mm) paper
honeycomb covered with Melanex which would
give an estimated weight of 5 lbm [2.3 kg]. He
rejected, after consideration, the idea of using the
shell as the principal load-carrying member, and
he employed a fairly conventional tubular "spine"
frame. He decided to avoid the problems of wind-
screen fogging by leaving the rider's head in the
open: ". . .no bicyclist would want to be hermeti-
cally sealed in, or object to the sun, wind or rain
on his face in moderation."

Towing tests were made to determine drag,
and it was estimated that a touring bicyclist might
increase his average speed (without stops) by up
to 6 miles/h [2.68 m/sec].

Rowable bicycle

Kazimierz Borkowski was another entrant who
constructed a prototype. His is a machine propelled
by a sliding-seat action along the very long cross
bar (Figure 12.4). The seat is attached to a carriage
which, during the power (backward) stroke, engages
the long loop of chain coming from the rear wheel.
The handlebars do not move longitudinally, so
that the rider must alter his position considerably
during the stroke.

Mr. Borkowski claims no more than that this is a
"sport and recreation" vehicle, and that it gives
healthy exercise to more muscles in the body than

Figure 12.3
Lydiard "Bicar"—Mark IV.
Lydiard's proposed further
development of his Bicar
would have sprung wheels
with pull rods operating
possibly a variable-ratio
friction gear in the rear
wheel. From reference 3.

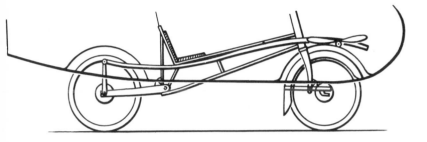

Figure 12.4
Borkowski's rowing-action
bicycle. This machine is
driven through a sliding
seat which runs on a long
crossbar, power being
transmitted on the
backward stroke.
From reference 3.

is the case for normal cycling. The judges were obviously concerned with the change in attitude and center of gravity of the rider.

Semienclosed recumbent bicycle

Stanislaw Garbien's vehicle included power transmission to the rear wheel through swinging constant-velocity cranks and an infinitely variable gear. His machine is a bicycle in which the rider sits fairly high up over the rear wheel and pushes levers over the front wheel (Figure 12.5). To enable the rider to put his feet on the ground when starting and stopping, the machine has an open-sided body.

The hopeful future

It does not take a great stretch of the imagination to visualize improved bicycles, whether of these designs or of others yet untried, being used for city-wide mass transportation. Yet such a view may be too fanciful. Vehicles powered by human muscle power alone are not going to be welcomed by all. There have been some concepts, and some prototype developments, of transportation systems based on bicycles or pedal-powered cars which can be attached to a powered guideway or cycleway (incorporating, for instance, a moving-cable towing system) for steep hills or long stretches[4-7] (see Figures 12.6 and 12.7) which may be more acceptable for the aged and the less-energetic among us.

Whether any of these seemingly optimistic developments will actually take place, or whether the world will continue to rush to utilize every new discovery of stored energy in ever-more-extravagant "power trips," cannot be predicted. What can be forecast is that the pattern of doubling energy consumption every decade or so cannot continue for much longer for many reasons, of which the limited availability of energy is only one. Pollution levels, land-use problems, and the availability of materials from which to make all the energy-using gadgets which this increasing consumption presupposes, are almost immediate problems in several countries. And world-wide, man's energy dissipation, presently about 1/30,000th of the incident

Figure 12.5
The Garbien semienclosed bicycle. Springing of both wheels is provided in this design. Power to the rear wheel is through swinging constant-velocity cranks and an infinitely variable gear. From reference 3.

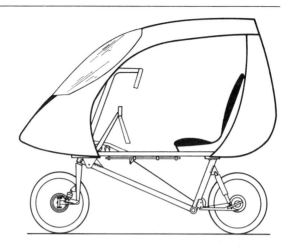

Figure 12.6
Transportation system for cyclecars.
From reference 6.

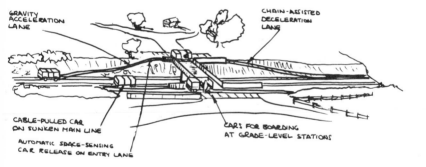

solar energy, would reach the same level as the sun's warmth on earth in about 115 years if we continued the present rate of increase. Obviously long before this condition could occur the climate would be so modified as to make irreversible changes in the whole of earth's ecology, and probably life would be impossible for many plants and creatures.

The gentle way of the bicycle for short distances and of the cycleway for somewhat longer journeys are transportation alternatives which are compatible both with nature and with a way of life which many would find an improvement over today's frenetic rushing hither and thither. We believe that the present renewed enthusiasm for bicycling is an encouraging sign of a saner future.

Figure 12.7
The Syracuse Crusway powered-guideway concept. Courtesy of Syracuse University Research Corporation.

References
Chapter 12

1. J. Y. Harrison et al., "Maximizing human power output by suitable selection of motion cycle and load," *Human Factors,* vol. 12, no. 3, 1970, pp. 315-329.

2. David Gordon Wilson, "A plan to encourage improvements in man-powered transport," *Engineering* (London), vol. 204, no. 5283, 21 July 1967, pp. 97-98.

3. David Gordon Wilson, "Man-powered land transport," *Engineering* (London), vol. 2071, no. 5372, 11 April 1969, pp. 567-573.

4. David Gordon Wilson, editor, "Personal-transit dual-mode cable cars," report DSR 72813-1, Engineering Projects Laboratory, Massachusetts Institute of Technology, Cambridge, Massachusetts, May 1971.

5. David Gordon Wilson, "Pallet systems for integrating urban transportation," *Transportation Engineering Journal,* American Society of Civil Engineers, vol. 98, no. TE 2, May 1972, pp. 225-242.

6. David Gordon Wilson, editor, "A research study of innovative transportation for new communities in Puerto Rico," Urban Systems Laboratory, Massachusetts Institute of Technology, Cambridge, Massachusetts, April 1971.

7. Andrew M. Valaas, "Archimedean-screw accelerators for automatic transportation," Mechanical engineering department, Massachusetts Institute of Techology, Cambridge, Massachusetts, June 1972.

Appendix: Some bicycle calculations

The following examples supplement those given in the text. They are intended to show how very simple mathematical models can yield valid predictions.

Downhill speeds

What speed is reached by a bicyclist free-wheeling down a slope? Assume that the rider and machine weigh 170 lbf [77 kg] and that they are on a long 5 percent (1 in 20) slope (see Figure A.1).

When the maximum speed is reached, the acceleration is by definition zero, and the force down the slope $(mg/g_c) \sin \alpha = 170/20$ lbf [37.8 newton] exactly balances the retarding forces of the rolling and wind resistances.

The relations for rolling resistance and wind resistance are taken from Chapters 5, 6, and an average frontal area of 3.65 sq ft [0.336 sq m] is assumed.

Assume that the rolling resistance is 11.5 lbf per long ton [0.0504 newtons/kg]. Then

$$\text{rolling resistance} = \frac{11.5 \text{ lbf/long ton} \times 170 \text{ lbf}}{2240 \text{ lbf/long ton}}$$

$$= 0.87 \text{ lbf [3.86 newtons]}.$$

The wind resistance is given by

$$\text{wind resistance} = 0.0023 \times \text{frontal area (sq ft)} \times [v(\text{mile/h})]^2.$$

Now the force down the slope can be set equal to the sum of the resistances, and the velocity v can be calculated:

$$\frac{170}{20} = 0.87 + 0.0023 \times 3.65 \times v^2;$$

$$v = \sqrt{\frac{8.5 - 0.87}{0.0084}} = 30.1 \text{ mile/h [13.5 m/sec]}.$$

On a 10 percent grade (1 in 10) the speed is 43.8 mile/h [19.6 m/sec] and on a 2.5 percent grade (1 in 40) it is 20.1 mile/h [9.0 m/sec]. These terminal speeds are reached asymptotically and therefore require an infinite distance to achieve. However, 95 percent of these terminal velocities would be reached in about a quarter mile (about 400 meters).

It is also of interest to investigate the reason why tandem bicycles run faster down hills than singles, a fact well appreciated by experienced bicyclists who have tried to keep up with tandem riders.

Record times for tandems, when compared with singles, show that although they are faster, they are so to a lesser degree than might be expected. The wind resistance of a tandem has been found to be about 30 percent greater than that of a single bicycle. The rolling resistance and force of gravity down the slope have been taken as twice those for a single bicycle.

A new calculation can now be carried out to find the speed of a tandem bicycle down a five-percent slope under similar conditions as those assumed for the single bicycle.

Figure A.1
Bicycle on a downhill slope.

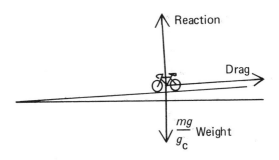

Rolling resistance + wind resistance

0.87 lbf $\times$ 2   + 0.0023 $\times$ 3.65 $\times$ 1.3 $[v \text{ (mile/h)}]^2$

$$= \frac{170}{20} \times 2 \text{ lbf,}$$

1.74 lbf   + 0.0109 $[v \text{ (mile/h)}]^2$ lbf

$$= 17 \text{ lbf,}$$

$$v = \sqrt{\frac{17 \text{ lbf} - 1.74 \text{ lbf}}{0.0109 \text{ lbf}}} \ \frac{\text{mile}}{\text{h}} = 37.4 \text{ mile/h}$$
$$[16.7 \text{ m/sec}].$$

The coefficients of wind and rolling resistance quoted are associated with only the "fastest" machines on very good surfaces. The calculations, however, give credence to reports of speeds of over 50 mile/h [22.4 m/sec] by riders in the Tour de France and other races in mountainous courses.

Power required for hill climbing

This calculation is included to show how great is the opposition to movement of bicyclists caused by gradients. Find the horsepower, the pedaling rate, and the pedaling thrust of a bicyclist proceeding at 9 mile/h [4.02 m/sec] up a hill of gradient 1 in 30 (3.33 percent). The mass of the man plus machine is 180 lbm [81.5 kg] and it is assumed that the resistance to motion caused by wheel rolling and air friction is 2 lbf [8.9 newtons]. The machine is geared to 65 in. [5.2 m] (the movement for one crank revolution is $65\pi/12$ or 17 feet) and the crank length is 6½ in. [165 mm].

The component of weight down the slope is

$$\frac{180 \, g}{30 \, g_c} = 6 \text{ lbf [2.66 newtons]}$$

The total force to be overcome is
2 lbf + 6 lbf = 8 lbf [3.56 newtons]

Hence power output is

$$9 \text{ mile/h} \times \frac{88 \text{ ft/sec}}{60 \text{ mile/h}} \times \frac{8 \text{ lbf}}{550 \text{ (ft lbf/sec)/hp}}$$
$$= 0.192 \text{ hp [144 watts]}$$

The pedaling rate is

$$\frac{13.2 \text{ ft/sec}}{17 \text{ ft}} \times 60 \text{ sec/min} = 46.6 \text{ rpm}$$

In one crank revolution the bicycle moves 17 ft against a force of 8 lbf; hence work done is 17 ft × 8 lbf = 136 ft lbf [184 joules]

In one crank revolution the bicyclists moves each foot through

$$\frac{2 \times 6\frac{1}{2} \text{ in.} \times \pi}{12 \text{ in./ft}} = 3.4 \text{ ft } [1.036 \text{ m}]$$

Hence the mean pedal thrust P is

$$\frac{136 \text{ ft lbf}}{3.4 \text{ ft}} = 40 \text{ lbf } [178 \text{ newtons}]$$

This assumes no pedal pull on the upstroke and 100 percent transmission efficiencies. The power output, at the given pedaling rate, has been shown to be perfectly feasible for many young men when pedaling on ergometers for periods of one-quarter of an hour.

Sharp[1] gives details of work by R. P. Scott in 1889[2] on measuring the actual pedal thrusts exerted by riders under various conditions. A particular example concerns the movement of a "rear driver geared to 54 in. [4.2 m] up a gradient of 1 in 20 (5 percent) at 9 miles per hour [4 m/sec]" which set of circumstances is similar to those assumed for the above calculation. The pedal thrust was shown to vary greatly, ranging from near zero to 150 lbf [665 newton], during the pedal revolutions.

In order to investigate the above phenomenon the senior author constructed an ergometer fitted with a calibrated braking device so that power input could be measured.[3] In addition a type of compressible pedal similar to that used by R. P. Scott was used. The compression of this pedal caused the movement, via a lever system, of a pen which traced the variation of the pedaler's thrust on a moving paper band, driven by the crank-set. Experimental results using the ergometer showed that when pedaling at about 60 rev/min and producing about 0.1 hp [74.6 watts] or less the author (FRW) could so skillfully move his ankles that the average applied thrust to the pedal was about 1.4 times the average tangential thrust required. If a straight up-and-down thrust were assumed the ratio would be 1.66.

However when the author tried to develop higher power outputs by increasing his pedaling rate, the above-mentioned ratio rose gradually to about 2.5 at a power output of 0.35 hp [262 watts] It appears that prolonged practice in pedaling at high power outputs may fit a racing man to economize in effort. "Getting in the miles" is common advice given to the racing man. The basis of this advice may lie in prolonged practice being necessary for efficient pedaling at the high rates and foot thrusts involved (see Figure 2.2). It is known that competent racing bicyclists can show better oxygen-usage efficiencies when pedaling ergometers than do other athletes unaccustomed to pedaling crank-sets, and thus adding evidence of the need for pedaling practice.

Riding around curves

The following calculations are included in order to show the reasons for the use of banking on roads and racing tracks.

Determine the speed at which a bicycle would commence to slide tangentially when traveling on a flat surface and rounding a bend of 100 ft [30.48 m] radius r if the coefficient of friction μ between tires and road is assumed to be 0.6.

At the point of skidding, the turning force necessary to give the bicycle the inward acceleration necessary for it to negotiate the bend is just equal to the maximum grip of the tires on the road,

$$\frac{mv^2}{rg_c} = \mu \frac{mg}{g_c}.$$

Hence

$$v = \sqrt{r\,\mu\,g} = \sqrt{\text{radius} \times \text{coefficient of friction} \times g}$$
$$= \sqrt{100 \text{ ft} \times 0.6 \times 32.2 \text{ ft/sec}^2}$$

= 44 ft/sec

= 30 mile/h [13.4 m/sec] .

How should the track be banked so that there is no tendency to skid at 30 mile/h?

The angle of bank of the track should be equal to the angle of bank of the bicycle at 44 ft/sec [13.4 m/sec] on a bend of 100 ft [30.48 m] radius. From the equilibrium diagram, (Figure A.2),

$$\tan \alpha = \frac{mv^2}{rg_c} \bigg/ \frac{mg}{g_c} = \frac{v^2}{rg} = \frac{(44 \text{ ft/sec})^2}{32.2 \text{ ft/sec}^2 \times 100 \text{ ft}}$$

$$= 0.6,$$

$$\alpha \cong 31 \text{ degrees.}$$

If it is assumed that the bicycle is running on this banked track at a speed higher than 30 mile/h [13.4 m/sec] , at what speed can the vehicle get round the bend without skidding?

The following relation can be shown to apply:

$$\text{Maximum speed } v = \sqrt{gr \frac{(\mu + \tan \alpha)}{(1 - \mu \tan \alpha)}}$$

$$= \sqrt{g \times \text{radius} \times \frac{\begin{array}{c}\text{(coefficient of friction} \\ + \text{ tangent of angle)}\end{array}}{\begin{array}{c}(1 - \text{ coefficient of friction} \\ \times \text{ tangent of angle)}\end{array}}}$$

Figure A.2
Bicycle on banked track in equilibrium.

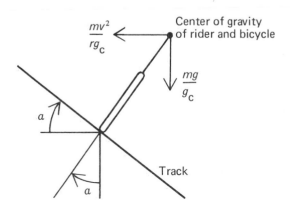

Substituting numerical values, with the coefficient of friction μ at 0.6, we have

$$v = \sqrt{32.2 \text{ ft/sec}^2 \times 100 \text{ ft } \frac{(0.6 + 0.6)}{(1 - 0.6 \times 0.6)}}$$

$$= 78 \text{ ft/sec}$$

$$= 53 \text{ mile/h } [23.7 \text{ m/sec}].$$

The relatively simple calculations given above show that it is possible to estimate safe banking angles for all speeds and radii of tracks or roads. In practice other matters must be taken into consideration in connection with track architecture. Most bicycle tracks are small enough to result in large relative differences between the inner and outer-edge radii. As a consequence the banking at the outer edge can be less than at the inner. In the case of tracks for racing cars the size is generally much greater than that common for bicycles and the banking generally is made steeper at the outside edges than at the inner for reasons of safety.

When a bicycle rider travels round a bend he leans over at the equilibrium angle calculated above. With two-track vehicles no leaning is possible if the center of gravity is low enough—that is, they will skid rather than overturn. In the case of man-propelled tricycles at speed round bends, however, great contortions on the part of the rider are necessary to avoid overturning. Tricycles are also most difficult to ride round banked tracks at low speeds because of the strain of needing constantly to steer up the banking.

Tube materials and dimensions

Effect of tube-wall thickness: These examples are included in order to bring out the considerable effect that changes in wall thickness, diameter and material of construction have on the rigidity of tubing when deflected by loading.

If the gauge of 1-in-diameter [25.4 mm] tubing is reduced from 18 gauge (0.048 in. [1.22 mm]) to 23 gauge (0.024 in. [0.61 mm]) what would be the increase in deflection of the end of a straight handlebar when pulled by the grips?

This problem may be modeled as an end-loaded cantilever as shown in Figure A.3. The deflection (δ) is proportional to the moment of inertia (I) of the beam (tube) section:

$$\delta = F\ell^3/3EI,$$

where F is the force, ℓ the length, and E the Young's modulus.

The moments of inertia of the two tubes are given by $I = \pi(D^4 - d^4)/64$ (Figure A.4):

18-gauge tube, $D^4 - d^4 : 1^4 - (1-0.096)^4$
$= 1 - 0.66$ or 0.34 in.^{4}.

23-gauge tube, $D^4 - d^4 : 1^4 - (1-0.048)^4$
$= 1 - 0.82$ or 0.18 in.^{4}.

The moments of inertia are in the ratio of about 2 to 1 and as a consequence the deflection of the

Figure A.3
Loading of a cantilever beam.

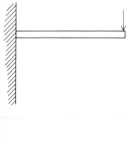

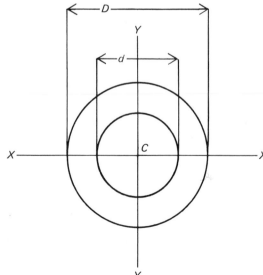

Figure A.4
Moment of inertia of a hollow tube.

$$I_{XX} = I_{YY} = \frac{\pi(D^4 - d^4)}{64}$$

18-gauge tube is about half that of the 23-gauge for the same loading.

Effect of tube material: If the material of construction is changed to one with a Young's modulus value of half, what effect will this have on the deflection of a given size of tube?

From the above formula it is seen that the deflection for a given set of circumstances is inversely proportional to the value of the Young's modulus (E). Hence the deflection is doubled. The ultimate strength of the material has no effect, of course, on the amount of deflection providing only that it is within the so-called "elastic" range.

Effect of tube diameter: If the tubing diameter is reduced from 1 in. [25.4 mm] to ½ in. [12.7 mm] and to keep the tube weight per unit length constant, the wall thickness is doubled, what effect will this have on the deflection if we assume 23 and 18 gauges, respectively?

The deflection is inversely proportional to the change in the moment of inertia (I) of the tubes, that is, to $(D^4 - d^4)$:

23-gauge 1-in.-dia tube:

$$(D^4 - d^4) = 1^4 - (1 - 0.048)^4$$
$$= 1 - 0.82 = 0.18 \text{ in.}^4$$

18-gauge ½-in.-dia tube:

$$(D^4 - d^4) = 0.5^4 - (0.5 - 0.096)^4$$
$$= 0.0625 - 0.0266 = 0.036 \text{ in.}^4$$

The moments of inertia I are in the ratio of about 4 to 1, and as a consequence the deflection of the 1-in. tube is only a quarter of that of the ½-in. tube.

$$(D^4 - d^4) = 0.5^4 - (0.5 - 0.096)^4$$
$$= 0.0625 - 0.0266 = 0.036 \text{ in.}^4$$

References
Appendix

1. A. Sharp, *Bicycles and tricycles* (London: Longmans, Green & Company, 1896), pp. 260-270.

2. R. P. Scott, *Cycling art, energy and locomotion* (J. B. Lippincott Company, 1889), pp. 48-59.

3. F. R. Whitt, "Ankling," *Bicycling,* February 1971, pp. 16-17.

Index